Materials Management in Sustainable Construction Engineering

Materials Management in Sustainable Construction Engineering

Edited by **Alistair Doyle**

WILLFORD PRESS

New York

Published by Willford Press,
118-35 Queens Blvd., Suite 400,
Forest Hills, NY 11375, USA
www.willfordpress.com

Materials Management in Sustainable Construction Engineering
Edited by Alistair Doyle

International Standard Book Number: 978-1-68285-054-1 (Hardback)

Printed in the United States of America.

Contents

Permissions

List of Contributors

Preface

Sustainable construction engineering focuses on reducing the negative impacts of construction on the environment. This process starts from the very beginning of construction like choosing recycled materials, efficient use of resources, reducing waste, etc. This book focuses on such practices, new technologies and methods for optimization. Materials efficiency, ecological design, waste management, integration of renewable energy systems, etc. are some of the topics discussed in this extensive text. This book will serve as a reference material for a broad spectrum of readers such as civil engineers, architects, environmentalists and students associated with this field.

All of the data presented henceforth, was collaborated in the wake of recent advancements in the field. The aim of this book is to present the diversified developments from across the globe in a comprehensible manner. The opinions expressed in each chapter belong solely to the contributing authors. Their interpretations of the topics are the integral part of this book, which I have carefully compiled for a better understanding of the readers.

At the end, I would like to thank all those who dedicated their time and efforts for the successful completion of this book. I also wish to convey my gratitude towards my friends and family who supported me at every step.

Editor

EVALUATION OF MAINTENANCE MANAGEMENT PRACTICE IN BANKING INDUSTRY IN LAGOS STATE, NIGERIA

Faremi Julius Olajide, Adenuga Olumide Afolarin

Department of Building, Faculty of Environmental Sciences, University of Lagos, Akoka, Lagos, Nigeria

*Corresponding E-mail : jydejulius@yahoo.com

ABSTRACT

The study focused on maintenance management practice in banking industry in Lagos state, Nigeria. It assessed the operational state of bank buildings, the factors affecting maintenance management of bank buildings, the maintenance management strategy used in maintaining bank buildings and the adequacy of funds available for maintenance management of bank buildings within the study area. In achieving these objectives, opinions of maintenance officers and users of randomly selected banks were sampled through structured questionnaires. The data collected were analyzed using descriptive and inferential statistics. The analysis revealed the operational state of bank buildings in Lagos State as good, there is no significant difference in the perception of the maintenance staff and users as to the operational state. However, there is significant difference in the operational state of the old and the new generation bank buildings as the study reveals that the components and services of the buildings of the new generation banks are in better operational state than those of the old generation bank buildings. Maintenance officers ranked attitude of users and misuse of facilities as the most significant factor affecting maintenance management of bank buildings while users of the buildings ranked lack of discernable maintenance culture in the country as the most significant factor responsible for poor maintenance management of bank buildings. The perception of the maintenance staff and users of bank buildings as regards factors responsible for poor maintenance management of bank buildings are significantly different. The study recommended proactive measures to keep hypothesized factors under check in other to overcome the prevailing maintenance problems of bank buildings. Top management are to provide adequate funding for the running of maintenance operations and such funds should be properly monitored to ensure that it is judiciously utilized.

Keywords: Bank Building, Funding, Maintenance, Maintenance Management, Maintenance Strategy and Operational State.

1.0 INTRODUCTION

Banks in Nigeria spend billions of naira in putting up buildings in order to cope with the exponential growth and expansion that has attended the industry in the last few years. Statistics shows that the Nigerian banking sector's asset base grew by approximately 227% between 2003 and 2007 [13]. These existing structures together with the structures to be erected in the nearest future, including that of the 716 micro finance banks already licensed create the needs for maintenance. Although much can be done at the design stage to reduce the amount of subsequent maintenance work, all elements of buildings deteriorate at a greater or lesser rate depending on materials, method of construction, age, environmental conditions, usage of building, and method of design and maintenance management of the building [1]. Assets such as buildings are a key resource for all types of organizations, including the banking sector. In the last few years, effects of the recapitalization of the banks in Nigeria has increased attention as the stock of new buildings costing billions of naira are being put into operation. This attention is due largely to the recognition of the significant contribution property makes to ultimate success or failure of a business and recognition of strategic importance of property to a company's financial structure.

However, in spite of millions of Naira spent to erect all these buildings, they are left as soon as commissioned to face premature but steady and rapid deterioration, decay and dilapidation.

Property resource, in the same way as human, financial and information resources contribute to the success of these organizations and need to be effectively and efficiently managed. These assets have to be professionally managed to ensure that the asset value is maintained

1.1 STATEMENT OF THE PROBLEM

Most of the bank buildings are in deplorable conditions of structural and decorative disrepair. Series of research have been carried out on factors responsible for the poor maintenance management of public, private housing estates and offices in Nigeria but only scant attention has been given to the evaluation of the maintenance management practices adopted in the implementation of maintenance programmes for banks and financial institutions buildings. There is therefore a need to establish and evaluate the strategies for the maintenance management practice of bank buildings using appropriate analysis.

1.2 AIM

To evaluate the maintenance management practice of the banking industry in Nigeria with a view to recommending most efficient maintenance management strategy.

1.3 OBJECTIVES

1. To assess the operational state (physical-functional condition) of bank buildings in Lagos state as carried out by the maintenance department.
2. To find out the maintenance management strategy used in maintaining bank buildings in Lagos State.

1.4 RESEARCH HYPOTHESES

➢ There is no significant difference in the perception of the maintenance staff and the users as to the operational state of bank buildings in Lagos state.
➢ There is no significant association between the maintenance strategy deployed by the maintenance department and the operational state of bank buildings in Lagos State.

2.0 LITERATURE REVIEW

It is highly desirable but hardly feasibly to produce maintenance free buildings. Much work can however be done at the design stage to reduce the amount of subsequent maintenance work that will be done during the operation and maintenance life of a building. [2] sees the main purposes of maintaining buildings as retaining its value of investment, maintaining the building in a condition in which it continues to fulfill its function, and presenting a good appearance. [6] developed a more specific definition of building maintenance as "The regular inspection of all parts of a building and the execution of work necessary to keep the structure, finishes and fittings in a proper and acceptable state of repair, including decoration, both internally and externally". [5] on their part see maintenance actions as technical and economic that tries to raise the quality level of a building element and/or restore it to the initial performance level in accordance with a certain requirement. [11] discloses that a prime maintenance is to preserve a building in its initial state, as far as practicable so that it effectively serves its purpose.

Building maintenance work is done to ensure that the building is in safe, healthy condition in accordance with specified standards. Maintenance of the built environment impacts on the whole nation. The conditions of the surroundings in which we live and learn, are a reflection of the nation's well being. 'The condition and quality of buildings reflects public pride or indifference, the level of prosperity in the area, social values and behavior and all the many influences both past and present which combine to give a community its unique character' [9].

An effective maintenance management system might be characterized as the product of prudence, of the sentiment that 'a stitch in time saves nine'. Good maintenance management systems are essential for economically viable and operationally safe buildings [10].

Historically, in both public and the private sectors, the maintenance is seen as an avoidable task which is perceived as adding little to the quality of the working environment, and expending scarce resources which would be better utilized [7]. The financial consequences of neglecting maintenance is often not only seen in terms of reduced asset life and premature replacement but also in increased operating cost and waste of related and natural and financial resources [3]. This is why property managers should give maintenance a high priority in their day to day activities [8].

Management of any process involves assessing performance, and maintenance management of buildings is no exception. In order for any maintenance manager to measure performance and set priorities, the organizational needs have to be considered i.e. the function and performance of buildings and their appropriate standards will be dependent on the user's perception and their primary needs [4]. Performance of bank buildings and their component depends to a large extent on continuous and planned periodical maintenance, which challenges owners and facility managers to institute precise planning based on a well-structured maintenance programmes [12]. Despite the ever-growing need for lower operational costs, facilities managers must ensure that facilities are constructed and maintained without compromising safety.

3.0   METHODOLOGY

This research covers bank buildings in Lagos State, Nigeria. A list of the twenty-two (22) banks with operations in Lagos State was obtained from the portal of the Central Bank of Nigeria. From this comprehensive list of banks, a selection of ten (10) banks was done using the simple random sampling method. The simple random sampling method was chosen so as to give equal chances to all the listed banks. Two categories of questionnaires were designed for this study and were directed to the maintenance staff and the users of these selected bank buildings respectively.

A total of eight (8) questionnaires were sent out to each of the ten selected banks, out of which four of the questionnaires were directed to the maintenance staff and four questionnaires were directed to the users of each of the ten bank buildings respectively. Thus a total of eighty (80) questionnaires were sent out to the ten selected banks of which a total of fifty-seven (57) questionnaires were completed and used for the analysis.

3.1   METHOD OF DATA ANALYSIS

The data collected through structured questionnaires were coded and assigned variables for easy handling through computer analysis using statistical package for social sciences (SPSS 15.0) so as to obtain a comprehensive and accurate analysis in both the descriptive and inferential statistics as applicable.

4.0   RESULTS AND DISCUSSIONS

Below are the analysis and the results of data collected from the field survey as extracted from the data collection instruments (Questionnaire A and B respectively).

Table 1: Number and rate of response by Maintenance Staff and Users

Category	Questionnaires sent out	Responses	% of Response
Maintenance staff	40 (50.0)	30 (75.0)	52.6
Users	40 (50.0)	27 (67.5)	47.4
Total	80	57	100.0

Table 2: Bank names and addresses (Maintenance staff and users)

Bank names	Number of questionnaire sent out		Total questionnaires received
	Maintenance Staff	Users	
UBA	4	4	6
WEMA	4	4	8
First Bank	4	4	6
Union	4	4	3
GTB	4	4	4
Stanbic IBTC	4	4	6
Skye	4	4	4
Access	4	4	6
Diamond	4	4	6
Zenith	4	4	7
Total	40	40	57

Source: (Field survey 2009)

Table 3: Category of respondent in the maintenance department

Category	Frequency	Percentage
Junior	15	51.7
Intermediate	14	48.3
Total	29	100.0

Source: (Field Survey 2009)

Table 3 shows a breakdown of the category of the respondents in the maintenance department, it reveals that 51.7% of the maintenance department staff are junior staff while 48.3% of the maintenance staff are at the intermediate category.

Table 4: The departments of maintenance staff in the bank

Department	Frequency	Percentage
Maintenance	13	43.3
Projects	5	16.7
Facilities	7	23.3
Premises and Property	5	16.7
Total	30	100.0

Source: (Field Survey 2009)

Table 4: shows the breakdown of the departments of the maintenance staff in the banks under study. The analysis shows that 43.3% of the respondents are in the maintenance department,

16.7% are in the projects department, 23.3% are in the Facilities department, 16.7% are in the Premises and property department.

Table 5: Experience of maintenance staff in the bank

Years of experience	Frequency	Percentage
Less than 5 years	24	80.0
6-10 years	6	20.0
Total	30	100.0

Source: (Field Survey 2009)

In table 5, the analysis shows that only 20% of the respondents have more than five (5) years experience of working as maintenance staff.

Table 6: Approximate number of full time employees in the maintenance department

Number of full time employees	Frequency	Percentage
1-5	1	3.3
6-10	11	36.7
11-15	9	30.0
16-20	6	20.0
21-25	3	10.0
Total	30	100.0

Source: (Field Survey 2009)

Table 6 shows the breakdown of the number of full time employees in the various maintenance departments of the banks in Lagos State.

Table 7: Maintenance staff perception of the condition of bank building

Building condition	Frequency	Percentage
Very poor	4	13.3
Poor	1	3.3
Good	10	33.3
Very good	15	50.0
Total	30	100.0

Source: (Field Survey 2009)

Table 7 shows that 13.3% and 3.3%, of the respondents in the maintenance department of the bank buildings in Lagos State perceived the overall condition of the buildings, as very poor and poor respectively.

Operational states of building elements and services as perceived by maintenance staff and the users of bank buildings in Lagos state, using the scale; (1) Very Bad (2) Bad (3) Average (4) Good (5) Very Good.

From the analysis, the following results were obtained;

Table 8: Operational state of building elements as perceived by users and maintenance staff of banks

Element	Mean	Remark
Structural elements (beams, columns, upper floor slabs and stairs)	4.6	Very good
Walls (external and internal walls)	4.26	Good
Finishes (wall finishes, floor finishes and ceilings)	3.83	Good
Widows	4.35	Good
Doors (external and internal doors)	4.03	Good
Roofs	4.33	Good
Services(sanitary appliances, building service equipment, disposal installation, water, ventilation, electrical, gas, lifts, protection installation, drainages, external services)	4.03	Good
Fittings and furniture	3.78	Good
Sanitation of the environment	3.93	Good

Source: (Field Survey 2009)

Table 9: The ranking of hypothesized factors responsible for poor maintenance management of bank buildings in Lagos state.

Hypothesized factors	Maintenance staff		Users	
	Mean	Rank	Mean	Rank
Attitude of users and misuse of facilities	4.30	1	4.70	2
Persistent breakdown through indiscipline and ignorance factors of building users	4.07	2	4.70	3
Insufficient fund for maintenance job	3.90	3	2.11	18
Natural deterioration due to age and environment	3.83	4	3.46	5
Inefficient inventory system	3.83	5	-	20
Inadequate training and development of personnel	3.53	6	3.15	6
Procurement of spare parts becomes difficult due to unavailable fund	3.50	7	2.11	17
Inflation of the cost of maintenance by the operatives	3.37	8	3.15	8
Lack of skilled manpower to maintain work in buildings designed and constructed by expatriates	3.25	9	2.93	10
Frequent shortage of materials and spare parts due to absence of efficient inventory system	3.14	10	2.93	11
Lack of skilled personnel in maintenance department	3.07	11	2.63	12
No effective maintenance due to de-emphasize in training, retraining and continue education	2.93	12	3.04	9
Use of poor quality component and materials	2.93	13	2.33	15

Hypothesized factors	Maintenance staff		Users	
	Mean	Rank	Mean	Rank
Inadequate/inappropriate maintenance of facility plant and equipment for maintenance operations	2.87	14	2.15	16
Complexity of design and non involvement of maintenance experts at the design stage	2.73	15	2.48	14
No long term arrangement for the supply of essential parts for replacement	2.73	16	3.15	7
Lack of successful maintenance programme by the maintenance department	2.67	17	2.00	19
Reluctance of some establishment to innovation supports	2.57	18	3.69	4
Absence of a form of planned maintenance programmes	2.47	19	2.63	13
Lack of discernable maintenance culture in the country	2.20	20	4.70	1

Source: (Field Survey 2009)

From the analysis, the maintenance staff rates the attitude of users and misuse of facilities as the most significant factors responsible for poor maintenance management of bank buildings.

On the contrary, the users rate lack of discernable maintenance culture as the most significant factor responsible for poor maintenance management of bank buildings.

In summary the following are the five most significant factors that affect maintenance management of bank buildings as ranked by the respondents:

Table 10: Top five hypothesized factors responsible for poor maintenance management of bank buildings in Lagos state.

Maintenance Staff	Users
Attitude of users and misuse of facilities	Lack of discernable maintenance culture in the country
Persistent breakdown through indiscipline and ignorance factors of building users	Attitude of users and misuse of facilities
Insufficient fund for maintenance job	Persistent breakdown through indiscipline and ignorance factors of building users
Natural deterioration due to age and environment	Reluctance of some establishment to innovation supports
Inefficient inventory system	Natural deterioration due to age and environment

4.1   TEST OF RESEARCH HYPOTHESES

Test of hypothesis 1

H_O = There is no significant difference in the perception of the maintenance staff and the users as to the operational state of bank buildings in Lagos State.

H_1 = There is significant difference in the perception of the maintenance staff and the users as to the operational state of bank buildings in Lagos State.

From the computation for hypothesis 1, the significance value for the **t** test was found to be **0.16.**

Decision: since **0.16>0.05** H_O (null hypothesis) is accepted.

Test of hypothesis 2

H_O = There is no significant association between the maintenance strategy deployed by the maintenance department and the operational state of bank buildings in Lagos State.

H_1 = There is significant association between the maintenance strategy deployed by the maintenance department and the operational state of bank buildings in Lagos State.

From the computation for hypothesis 4, the significance value for the Chi-square test was found to be **0.26.**

Decision: since **0.26>0.05** H_O (null hypothesis) is accepted.

5.0   CONCLUSIONS

The study showed that the operational state (physical-functional condition) of bank buildings in Lagos State was **good**, the mean been **4.0**. The analysis however revealed that there is significant difference in the operational state of the old generation banks when compared to the operational state of new generation bank buildings. The study revealed that the components and services of the buildings of the new generation banks are in better operational state than those of the old generation bank buildings.

As for the factors responsible for poor maintenance management of bank buildings in Lagos State, a number of hypothesized factors were identified with the degree of the significance of each of these factors established based on the responses obtained from the field survey as presented in the previous chapter of this research work. However, the two groups of respondents, the maintenance staff and the users have relative agreement on the degree of significance for all of the hypothesized factors except for the significance of lack of discernable maintenance culture in the country, which was ranked by the users as the most significant factor responsible for poor maintenance management of bank buildings whereas the maintenance staff ranked it as the least significant factor. Consequently, the research showed that the association between the perception of maintenance staff and users of bank buildings on the factors responsible for poor maintenance management is very strong. The information gathered in the course of this research work revealed that 97% of the maintenance departments of the bank buildings in Lagos State adopts wide managerial span of control as the organizational structure. This revelation depicts that the maintenance management of bank buildings in Lagos State has not given a chance to the use of the narrow span of control (ns alterative type of organizational structure) which may result in better coordination, supervision and monitoring of maintenance organization in banks and consequently a better overall performance of the department.

Furthermore, it was gathered that the maintenance departments of bank buildings deploy some form of preventive maintenance strategy, although the comprehensiveness and viability of such strategy was beyond the scope of this research work.

Finally, this research revealed that the level of funding of the maintenance department of banks is merely sufficient and therefore requires a thorough review by the top management to ensure that these buildings do not deteriorate prematurely resulting from poor or inadequate funding.

5.1   RECOMMENDATION

Strategies should be formulated by those saddled with the responsibility of maintenance management of bank buildings in line with systems and components of bank buildings to ensure that bank buildings remain operational in functional and safe manner. A viable preventive maintenance schedules (covering all major components and systems of bank buildings) should prepared and the implementation of such schedule should be absolute. This will foster pro-activeness in the maintenance management of systems and installations of the buildings and facilities in the banking industry, and will consequently preserve the huge investment of the banks in erecting the buildings and at the same time guarantee return on investment.

This study revealed that the funding of the maintenance department of bank buildings in Lagos State is **average** (with a mean value of 3.15). The adequacy of this level of funding for the maintenance needs of these buildings in the near future is doubtful as most of the buildings are relatively new at the moment with minimal breakdown arising from age and use of the facilities. Therefore, provision for better funding of the maintenance department should be made in subsequent years when systems and components of bank buildings begin to age and deteriorate at an increased rate. Such funds should however be monitored to ensure that they are judiciously utilized by the maintenance department.

Finally, maintenance management staff of banks should ensure that the listed hypothesized factors are kept within check as this will definitely assist them in planning and executing maintenance programmes, as well as overcome the prevailing maintenance problems of bank buildings.

REFERENCES

[1] Adenuga, O.A., Odusami, K.T. and Faremi, O.J. (2007) Assessment of Factors Affecting Maintenance Management of Public Hospital Buildings in Lagos State, Nigeria.

[2] Anderson, E.S and Jessen, S.A (2000) Project Evaluation Scheme: A Tool for Evaluation Project Status and Predicting Project Results, Project Management, 6(1), 61 – 69.

[3] Banful, E (2004) A Stitch in Time Saves Nine; Cultivating a Maintenance Culture in Ghana, An Article Presented at a Seminar on Maintenance Culture in Ghana. March 16, 1 – 2.

[4] Chanter, B. and Swallow, P. (2007) Building Maintenance Management, Blackwell Scientific, Oxford, England.

[5] Colen, I.F and De Brito, J (2002) In Vral, O, Abrantes, V. and Taden, A. (eds). Building Façade Maintenance Support System, 3, 1899 – 1907.

[6] Crips, D.J (1984) Building Maintenance.... A Client's Viewpoint, Managing Building Maintenance, 23 – 35, CIOB: London.

[7] Higher Education Backlog Maintenance Review (1998), London.

[8] Lai, J.H, Yik, F. W, Jones, P., (2009). Maintenance Cost of Chiller Plants in Hong Kong Building Services and Engineering Technology; 30: 65-78.

[9] Lee, R. (2001) Building Maintenance Management, Blackwell Science Ltd, Oxford: UK.

[10] Rapp, R. and George, B. (1998), "Maintenance management concepts in construction.

[11] Seeley, I.H. (1976) Building Maintenance, 5, Macmillian Press Ltd: London

[12] Shohet, I. Puterman M. and Gilboa E. (2002), Deterioration Patterns of Building Cladding Components for Maintenance Management.

[13] Soludo, C. (2008) "Guidelines and Incentives on Consolidation in The Nigeria Banking Industry". Press release - April 11, 2008 on Banking Sector Consolidation: Special Incentive to Encourage Weaker Banks.

APPLICATION OF ULTRA HIGH PERFORMANCE FIBER REINFORCED CONCRETE – THE MALAYSIA PERSPECTIVE

Yen Lei Voo[1], Behzad Nematollahi[2], Abu Bakar Bin Mohamed Said[3], Balamurugan A Gopal[3], Tet Shun Yee[4]

[1] Dura Technology Sdn. Bhd., Malaysia
[2] Civil Engineering Department, Universiti Putra Malaysia (UPM), Serdang, Malaysia
[3] Jabatan Kerja Raya, Kinta Daerah, Perak, Malaysia
[4] TS Yee and Associates, Perak, Malaysia

*Corresponding E-mail: dura@dura.com.my

ABSTRACT

One of the most significant breakthroughs in concrete technology at the end of the 20th century was the development of ultra-high performance fiber reinforced concrete (UHPFRC) with compressive strength and flexure strength beyond 160 MPa and 30 MPa, respectively; remarkable improvement in workability; durability resembled to natural rocks; ductility and toughness comparable to steel. While over the last two decades a tremendous amount of research works have been undertaken by academics and engineers worldwide, its use in the construction industry remain limited and it is particularly true in the Malaysian context. Aiming to utilizing the technology as an alternative for conventional solutions and within the vision of sustainable construction, it is the intent of this paper to demonstrate how UHPFRC can be used as both a sustainable and economic construction material. In general, UHPFRC structures are able to give immediate saving in terms of primary material consumption, embodied energy, CO_2 emissions and global warming potential. The major focus of this paper is to present both the various completed and on-going examples of UHPFRC application in Malaysia.

Keywords: *Ultra high performance, Fiber, Bridge, Retaining wall, Bridges*

1.0   INTRODUCTION

Remarkable development had been discovered during the last two decades in the field of concrete technology. One of the greatest breakthroughs was the development of ultra-high performance fiber reinforced concrete (UHPFRC).

What is UHPFRC? In short, Figure 14 shows UHPFRC belong to the group of High Performance Fiber Reinforced Cement Composites (HPFRCC), where HPFRCC defined as the kind of Fiber Reinforced Concretes (FRC) that exhibit strain-hardening under uniaxial tension force. In addition, UHPFRC is characterized by a dense matrix and consequently a very low permeability when compared to HPFRCC and normal strength concretes.

In Malaysia, UHPFRC was firstly introduced by Dura Technology Sdn. Bhd. in year 2007 with compressive strength and flexural strength of over 160MPa and 30MPa, respectively; however, it has only started its industrial-commercial penetration into the market as a new sustainable construction material since last 3 years. In general, UHPFRC is suitable for use in (i) the fabrication of precast elements for civil and structural engineering (such as bridge components), (ii) archi-structural features, (iii) durable components exposed to marine or aggressive environments, (iv) blast or impact protective structures, (v) strengthening material for repair/rehabilitation work for deteriorated reinforced concrete structures, (vi) portal frame building construction, and others. UHPFRC is a highly homogenous cementitious-based composite without coarse aggregates that can attain compressive strengths more than 160MPa. The standard mix design of UHPFRC is given in Table 7.

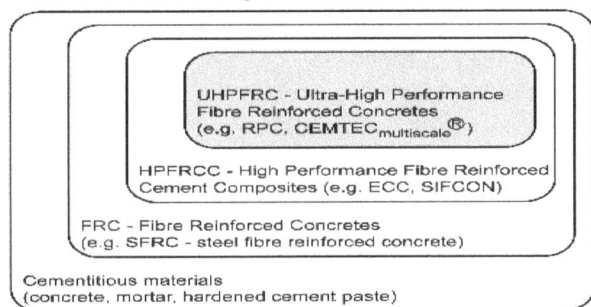

Figure 14: Classification of fiber reinforced concrete.

Table 7: Mix design of UHPFRC.

Ingredient	Mass (kg/m^3)
UHPC Premix	2100 – 2200
Superplasticizer	30 – 40
Steel Fiber	157
Free Water	144
3% Moisture	30
Targeted W/B Ratio	0.15
Total Air Void	< 4%

Table 8 summarizes the material characteristics of UHPdC and is compared against normal strength concrete (NSC) and high performance concrete (HPC). The comparison shows that UHPdC have superior mechanical properties over NSC and HPC in all aspects.

Table 8: Material characteristics of UHPdC compared to normal strength concrete (NSC) and high performance concrete (HPC)

Characteristics		Unit	Codes / Standards	NSC	HPC	UHPdC
Specific Density, ρ		kg/m^3	[1]	2300	2400	2350 – 2450
Cylinder Compressive Strength, f_{cy}		MPa	[2]	20 – 50	50 – 100	120 – 160
Cube Compressive Strength, f_{cc}		MPa	[3]	20 – 50	50 – 100	130 – 170
Creep Coefficient at 28 days, ϕ_{cc}			[4]	2 – 5	1 – 2	0.2 – 0.5
Post Cured Shrinkage		$\mu\varepsilon$	[4]	1000 – 2000	500 – 1000	< 100
Modulus of Elasticity, E_o		GPa	[5]	20 – 35	35 – 40	40 – 50
Poisson's Ratio, ν				0.2	0.2	0.18 – 0.2
Split Cyl. Cracking Strength, f_t		MPa	[6] or [7]	2 – 4	4 – 6	5 – 10
Split Cyl. Ultimate Strength, f_{sp}		MPa		2 – 4	4 – 6	10 – 18
Flexural 1st Cracking Strength, $f_{cr,4P}$		MPa	[8] (Four-Point Test on Un-notched Specimen)	2.5 – 4	4 – 8	8 – 9.3
Modulus of Rupture, $f_{cf,4P}$		MPa		2.5 – 4	4 – 8	18 – 35
Bending Fracture Energy, $G_{f,\delta=0.46mm}$		N/mm		< 0.1	< 0.2	1 – 2.5
Bending Fracture Energy, $G_{f,\delta=3.0mm}$		N/mm		< 0.1	< 0.2	10 – 20
Bending Fracture Energy, $G_{f,\delta=10mm}$		N/mm		< 0.1	< 0.2	15 – 30
Toughness Indexes	I_5			1	1	4 – 6
	I_{10}			1	1	10 – 15
	I_{20}			1	1	20 – 35
Rapid Chloride Permeability		coulomb	[9]	2000 – 4000	500 – 1000	< 200
Chloride Diffusion Coefficient, D		mm^2/s	[10]	$4 – 8 \times 10^{-6}$	$1 – 4 \times 10^{-6}$	$0.05 – 0.1 \times 10^{-6}$
Carbonation Depth		mm	[11]	5 – 15	1 – 2	< 0.1
Abrasion Resistance		mm	[12]	0.8 – 1.0	0.5 – 0.8	< 0.03
Water Absorption		%	[13]	> 3	1.5 – 3.0	< 0.2

2.0   APPLICATIONS

This paper gives an overview on the detail on some of the successful examples and on-going projects on the application of UHPFRC technology in Malaysia.

To-date, many prototype UHPFRC structures have been constructed in various countries such as France, USA, Germany, Canada, Japan, South Korea, Australia, New-Zealand, and Malaysia. According to Adeline et al. [14], the first application of UHPFRC was the UHPFRC in-filled steel tube composite used in the construction of a footbridge in 1997 at Sherbrooke, Canada. Since then, UHPFRC has caught the attention of academics, engineers and many governmental departments worldwide. Deem [15] reported that the first fully UHPFRC footbridge spanning 120 meters in the world was constructed in Seoul, South Korea in 2002. Subsequently, a motorway bridge was designed by VSL (Australia) at Shepherds Gully Creek, Australia, and was opened to traffic in 2005 [16]. According to Graybeal [17], UHPFRC can be used in a broad range of highway infrastructure applications due to its high compressive and tensile strengths and its enhanced durability properties; thereby allowing a longer design/service life and thin overlays, claddings, or shells. In addition, UHPFRC is also being considered to be used in a range of other applications such as precast concrete piles [18], seismic retrofit of substandard bridge substructures [19, 20], thin-bonded overlays on deteriorated bridge decks [21], and security and blast mitigation applications [22, 23]. Accomplished bridge projects using UHPFRC such as Sherbrooke footbridge (in Canada), Seonyu footbridge (in South Korea), Bourg-Les-Valence Bridge (in France), and Shepherds Gully Creek Bridge (in Australia) emphasize the high capability of UHPFRC to be used in infrastructural projects [24]. Figure 2 presents a schematic drawing showing the evolution of UHPdC technology with respect to structural and architectural applications from 1995 to 2010.

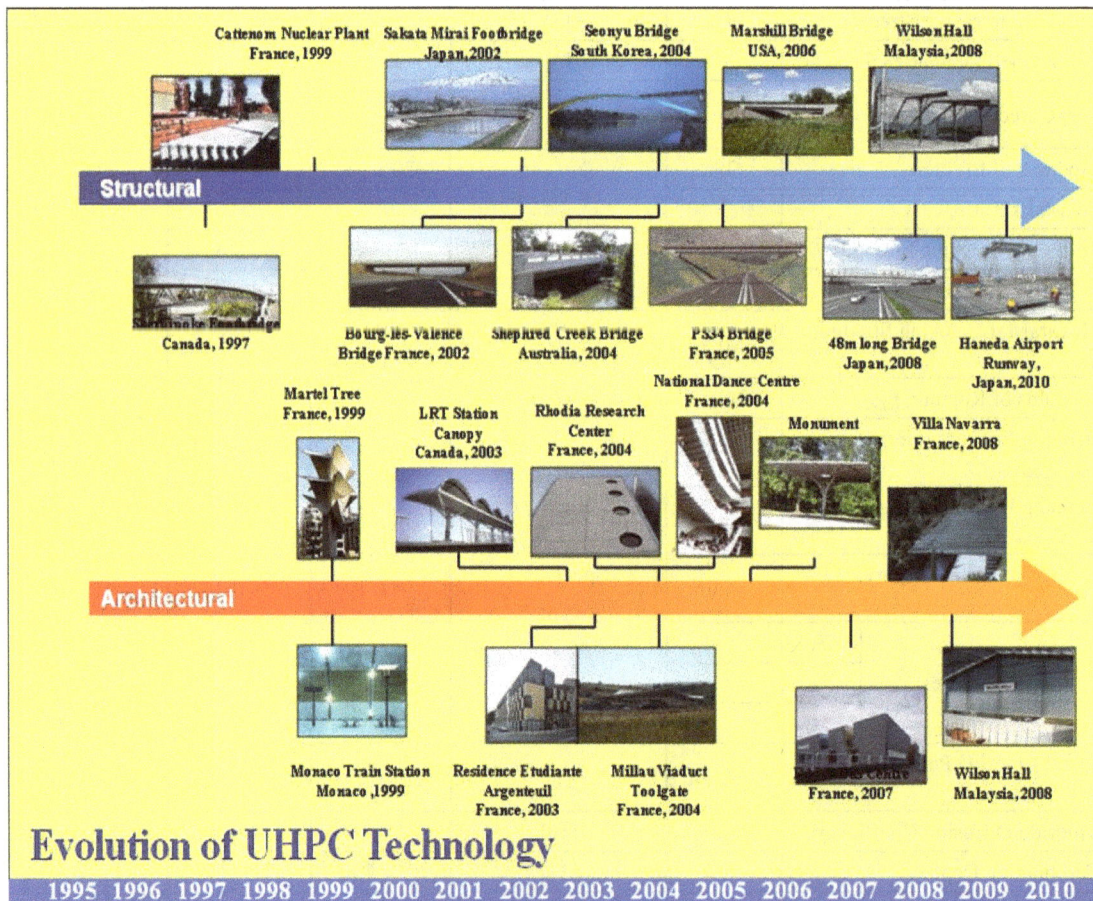

Figure 15: Application of UHPFRC.

2.1 UHPFRC PORTAL FRAME BUILDING (WAREHOUSE SOLUTION), COMPLETED

In year 2008, a portal frame building named *Wilson Hall* with a roof coverage area of 2,861m^2 was built using the prefabricated system of UHPFRC technology. The total transverse width and longitudinal length of the building is 67 m and 42.7 m, respectively. Each UHPFRC portal frame was spaced at 12.2 m c/c and the building consists of eight pieces of UHPFRC prestressed columns, internal rafters, cantilever rafters and connections as shown in Figure 16a. Details of the R&D works and construction sequences of the building can be obtained from Voo and Poon [25]. To-date, this building is the world first attempt to replace conventional steel beam with the UHPFRC prestressed beams/columns. This building has earned a place in the *Malaysia Book of Records* in year 2010. Besides that the paper Voo and Poon [25] also won the *JCI-OWICS Award 2008* where the paper was recognized as the most outstanding and original paper at the International Conference on Our world in Concrete & Structures.

Figure 16b presents the environmental impact calculation (EIC) of the UHPFRC portal frame system against the conventional steel portal frame system. In terms of material consumption, the UHPFRC portal frame system consumed 13% less material than the conventional steel portal frame solution. With regard to immediate construction cost, the UHPFRC system is 16% more economical that the conventional steel structural system. More cost savings can be realised for those factory buildings that are located in corrosive environment or places constantly subjected to chemical attack such as chemical plants, due to conventional steel structure would require periodic maintenance. In terms of environmental indexes, the UHPFRC solution has 24% less embodied energy and 19% less CO_2 emissions. For the 100-year GWP, the UHPFRC solution provides a reduction of 16% to the conventional solution. Thus this shows that the UHPFRC system can give a more sustainable solution against the conventional method.

(a)

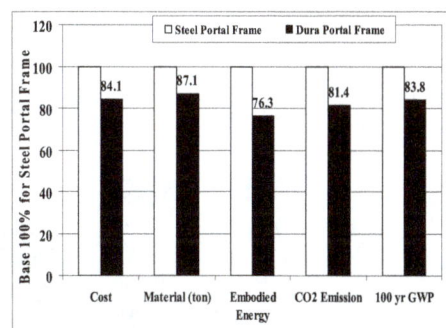

(b)

(c)

Figure 16: (a) UHPFRC portal frame (before completion at year 2008), (b) environmental impact calculation and (c) completed *Wilson Hall* (photo taken at year 2012).

2.2   ULTRA-LIGHT WEIGHT WALL PANEL (SECURITY SOLUTION), COMPLETED

One of the remarkable properties of UHPFRC is, it is a highly workable (i.e. flowable) composite material and it has great self-compacting ability. Its superior mechanical properties such as flexural strength made it an ideal material for the manufacturing of thin and light-weight wall panels where conventional steel reinforcements are entirely removed from the panels. Unlike conventional RC wall panel, the UHPFRC wall panel has negligible concern about corrosion issue as conventional steel reinforcement is absent in any parts of the structure.

Figure 17a shows an example of a total of 56 m long free standing anti-climb protective wall panels that was installed at the *Wilson Hall*. The wall was constructed in year 2008. Each wall panel has a total height of 7 m and a total width of 2 m, and comes with a self-weight of 2400 kg per piece. The wall panel consists of thin wall panel of 30 mm in thickness, two ribbed beams as wide as 75 mm and a base pad of 100 mm in thickness (refer to Figure 17b).

The wall panel has multiple applications such as it can use as thin wall panel against wind/rain/sun-shine/dust/spy. Besides, the wall panel also serves as acoustic panel against noise; security or anti-climb panel against thief; protective panel against minor blast and impact loading; impermeable membrane against highly corrosive compound and fire. The benefits of the UHPFRC wall when compared to conventional RC wall is that it is highly durable and impermeable, thus suitable for use in extremely aggressive environments such as marine environments or chemically active plants. The wall is easy to install as simple conventional drop-in anchors or pre-positioned bolts and nuts are used to connect the wall panel to the floors (except some grout may be needed for uneven floor base). No scaffolding, props or formwork are required over the entire installation, thus reducing construction site activities, improving safety margins and eliminating in-situ casting work. Besides, it is many times lighter than conventional RC wall system.

Other advantages are that the wall is guaranteed to be geometrically stable as they are steam-cured to minimized creep and long-term shrinkage and in term of finishing, it is aesthetically pleasing as its finish surface is smooth. More details of the wall panel can be found in Poon et al. [26].

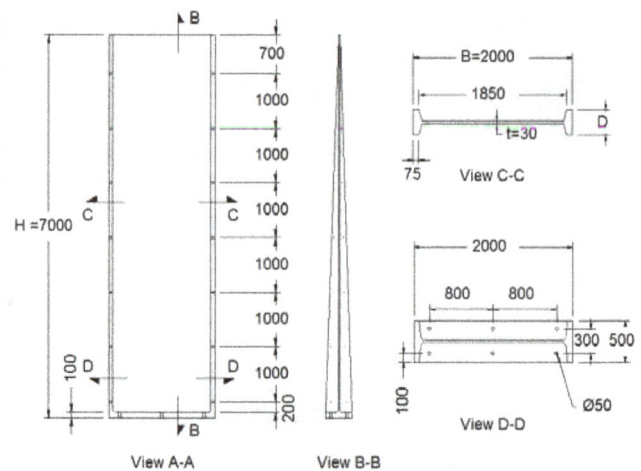

(a)   (b)

Figure 17: (a) UHPFRC anti-climb wall panel, (b) detail of UHPFRC anti-climb wall.

2.3 MONSOON DRAIN (HYDROLOGY SOLUTION), COMPLETED

Figure 18a shows a total of 180 m long by 1.5 m high retaining wall was used in the construction of a 90 m long monsoon drain for a housing development project in Ipoh, Perak. The L-shaped wall comes with thin panels of 30–50 mm thick (see Figure 18b). Unlike conventional RC L-shaped wall which is precast in a standard 1 m length and weights 1200 kg/m of wall, the UHPFRC retaining wall is made in 3 m lengths per piece (see Figure 18c) and has a self-weight of 260 kg/m, which gives a factor of five times lighter than the conventional solution. Prior-to construction of the wall, the local council (i.e. Majlis Bandaraya Ipoh) requested a load proof test on the wall with a surcharge load of 10 kPa at service and 15 kPa at ultimate. The wall was tested with back filled soil up to 1.5 m and an additional surcharge load of 25 kPa, which is 66% greater than the strength limit requirement and still it did not fail! Thus, the wall performance was deemed to satisfy with the design service and strength requirements.

Figure 18d shows a comparison of the EIC results of the UHPFRC retaining wall system against the conventional L-shaped RC wall as given in Figure 18c. In terms of material consumption, the UHPFRC retaining wall consumes 73% less material than the conventional RC wall. In terms of the environmental indexes, the UHPFRC wall requires less embodied energy and produces 49% less CO_2 emissions. In terms of the 100-years GWP, the UHPFRC solution provides a reduction of 43%. This it is another good example of how with innovative design UHPFRC technology supports sustainable construction solutions.

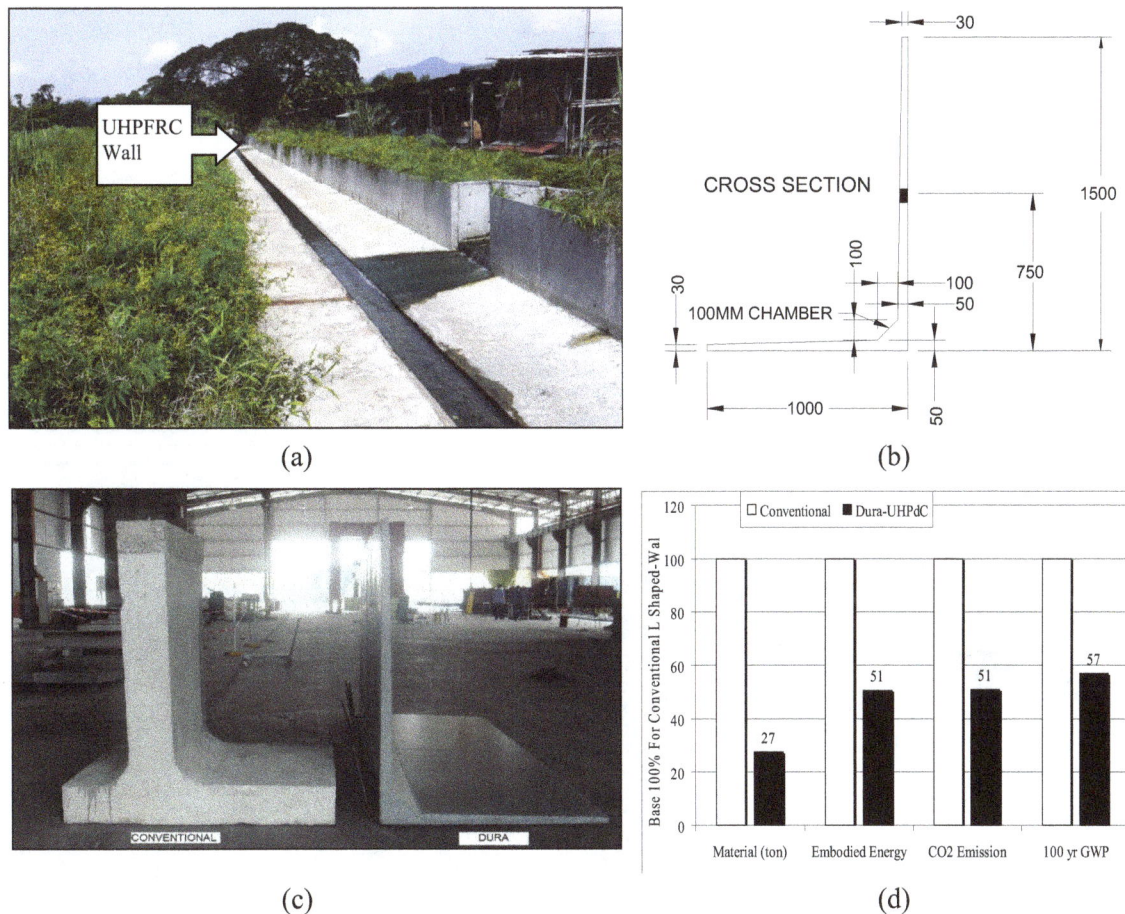

(a)

(b)

(c)

(d)

Figure 18: (a) 90m long monsoon drain using UHPFRC retaining wall, (b) cross-section detail; (c) comparison of conventional precast L-shape retaining wall against ultra-light weight UHPFRC retaining wall, and (d) EIC of UHPFRC retaining wall.

2.4 CANTILEVER RETAINING WALL (GEOTECHNICAL SOLUTION), COMPLETED

UHPFRC is ideal for short retaining wall construction (for H < 3m) due to its ultra-high strength-ultra-light-weight feature. Figure 19a gives an example on the detail of a 2.5 m tall UHPFRC wall. The L-shaped wall comes with a total height of 2.5 m and a total width of 2 m per piece. Each of the walls weighs 1200 kg (i.e. 600 kg/m). Unlike conventional RC wall, the UHPFRC wall does not have transverse reinforcements or crack control bars in any part of the concrete section. The only conventional steel reinforcement used is the major longitudinal reinforcements located at the ribbed beams (i.e the stem and the base) to resist the critical design moment effect resulted from the imposed loadings.

Figure 19b shows the prototype of the UHPFRC L-shaped retaining wall. In December 2010, the JKR Perak has constructed a 76 m long with 2.5 m tall retaining wall at Jalan Kota Bahru (Daerah Gopeng, Perak) using the above mentioned UHPFRC retaining wall and it took five working days to complete the entire construction work, which included site clearing work, preparation of the granular base, placing and assembling of the walls, and back filling of the earth. This exercise shows the UHPFRC retaining wall system is able to provide speedy construction solution.

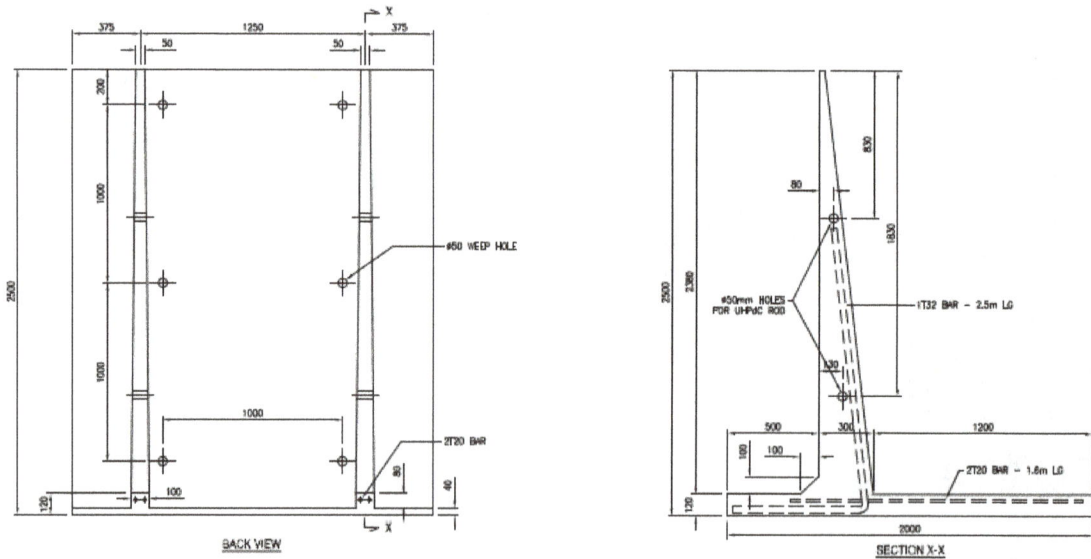

(a)

(b)

(c)

Figure 19: (a) Detail of 2.5 m high by 2 m wide UHPFRC retaining wall, (b) prototypes prior to transportation (back view), (c) 76 m long retaining wall installed at Jalan Kota Bahru, Gopeng, Perak.

2.5 50 M KAMPUNG LINSUM BRIDGE (MEDIUM TRAFFIC BRIDGE), COMPLETED

The JKR Negeri Sembilan was the first to use UHPFRC in the construction of a medium span motorway bridge at Kampung Linsum crossing a river call Sungai Linggi (see

Figure 20). The road bridge was completed in January 2011. To date, this bridge is the first in Malaysia and may also be the world's longest composite road bridge made from UHPFRC. The bridge was constructed using a single U-trough girder 1.75 m deep, 2.5 m wide at the top, topped with a 4 m wide cast in-situ reinforced concrete deck 200 mm thick. The UHPFRC girder ends were encased in normal strength concrete abutments at the bridge site and made integral with the abutment seating. The girder was built without any conventional shear reinforcement as the UHPFRC had considerable shear capacity. The UHPFRC used has achieved up to 180 MPa of compressive strength and 30 MPa of flexural strength. The bridge has also earned a status in the *Malaysia Book of Records* in year 2011. Detail of the construction of the composite bridge can be found in Voo et al. [27].

The precast girder consists of a total of seven segments, which consists of five standard internal segments (IS) each 8 m long that weighed 18 tons, and two end standard segments (ES) each 5 m long that weighed 15 tons (see

Figure 21). Unlike conventional precast concrete girders, the UHPdC girder does not have vertical shear link in its thin webs. The only conventional reinforcements used are the bursting reinforcement at the anchorage zone, lifting reinforcement at the tendon deflector positions, and horizontal shear reinforcement at the top flanges where connection with the RC deck is required.

Figure 20: Kampung Linsum Bridge, Rantau, Negeri Sembilan.

Initially the engineers who were engaged to design the bridge had proposed using two steel structural welded beams (see Figure 22a). Later on, the consultants chose to go with the UHPFRC girder design due to convincing argument and benefit of adopting an UHPFRC composite bridge design solution. Such benefits include no piers at the waterway of the river, much lower maintenance, more eco-friendly, better aesthetically and, most importantly, it was cheaper!

Figure 22b summaries the comparison of the EIC results between the UHPRFC and steel composite bridges. In terms of material consumption, the UHPFRC solution consumed 14% more material (in terms of weight) than the steel-composite girder solution. In terms of environmental impact, however, the UHPFRC solution had 66% less embodied energy and 57% less CO_2 emissions. In terms of the 100-year GWP, the UHPFRC solution gives a reduction of 52% over the steel-composite girder design. In addition to the environmental cost savings, the UHPFRC composite bridge superstructure resulted in a projected cost saving of 27%. Thus, the UHPFRC solution was not just better for the environment, it was a more economical solution based on initial costs. When maintenance costs are considered, the UHPFRC solution is vastly more economical!

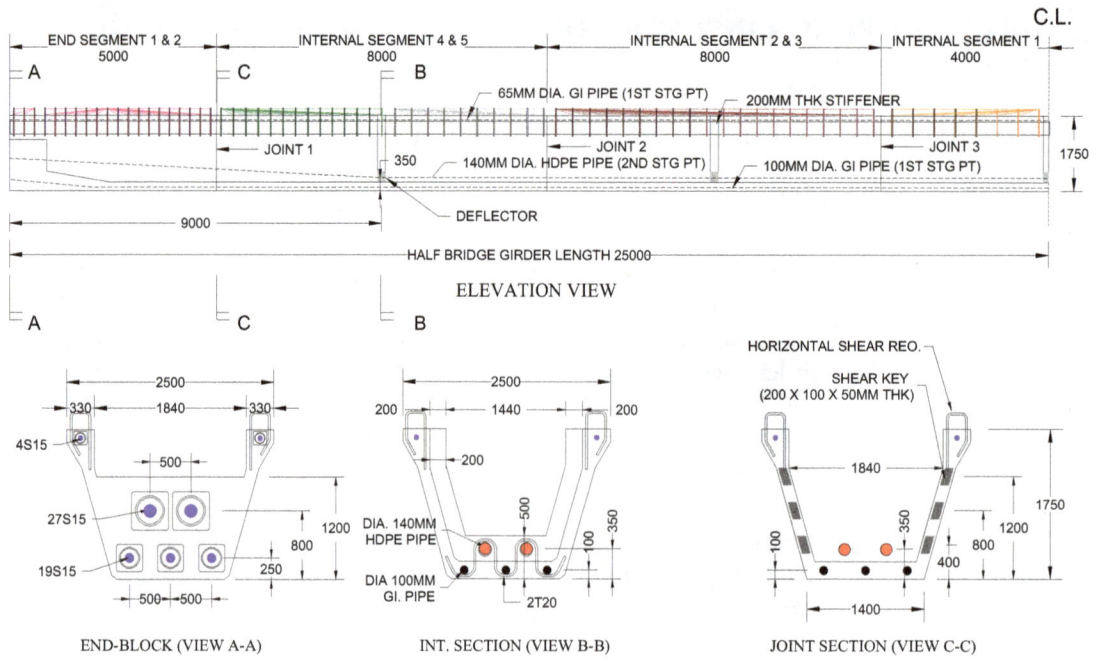

Figure 21: Detail of UHPFRC UBG1750 girder.

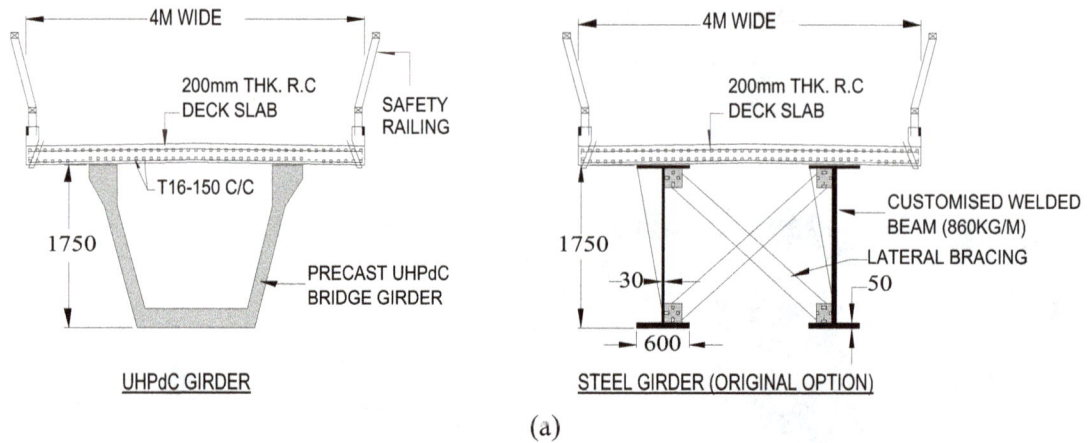

(a)

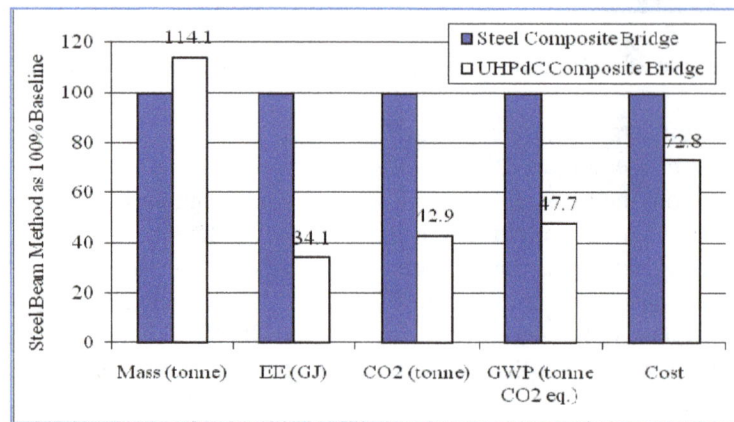

(b)

Figure 22: (a) Comparison of steel composite bridge against UHPFRC composite bridge, (b) EIC assessment (details in Voo et al. [3]).

2.6 25 M KAMPUNG ULU GEROH BRIDGE (MEDIUM TRAFFIC BRIDGE), COMPLETED

The JKR Perak (Kinta Daerah) was the first to use UHPFRC in Perak state in the construction of a short span motorway bridge at a small village called Kampung Ulu Geroh crossing a river call Sungai Itik (refer to

Figure **23**). To-date, this bridge is Malaysia first full UHPFRC bridge/deck system where the superstructure of the bridge is constructed without conventional RC deck. This bridge was designed to withstand 30 units HB loading and HA + KEL loading as per BD37/01. Construction of the bridge commenced at mid November 2011, and the bridge work was completed in mid January 2012 (which gives a construction period of 2 months). This bridge has a single span length of 25m and was constructed using two precast UHPFRC T-girders 1.375 m deep, 1.5 m wide at the top flange (refer to

Figure **23**).

The major obstacle in this project was the poor existing access road to the job site. The largest vehicles able to access to the site were the 20 tonnes capacity mobile crane and those ten wheels trucks which come with a tray length not exceeding 8 m. Given such constraint, the conventional precast RC beams was immediately ruled out in the design due to the self-weight of the 25 m long conventional precast RC beams which exceed the maximum possible carrying capacity of the two mobile cranes. The other possible option is using steel bridge where weight is not a major issue. However, the authority rule out this option too because maintenance is something they wanted to avoid. Besides, no centre pier is allowed in the waterway of the river. With these limitations, the UHPFRC bridge system proved to be the best solution as a single UHPFRC T-beam weighted only 25 tonnes and in addition the girder has remarkable durability. The UHPFRC girder ends were encased in normal strength concrete abutments and made integral with the abutment seating. Unlike any conventional concrete beam, the UHPFRC girders were built without any conventional shear reinforcement as the UHPFRC had considerable shear capacity.

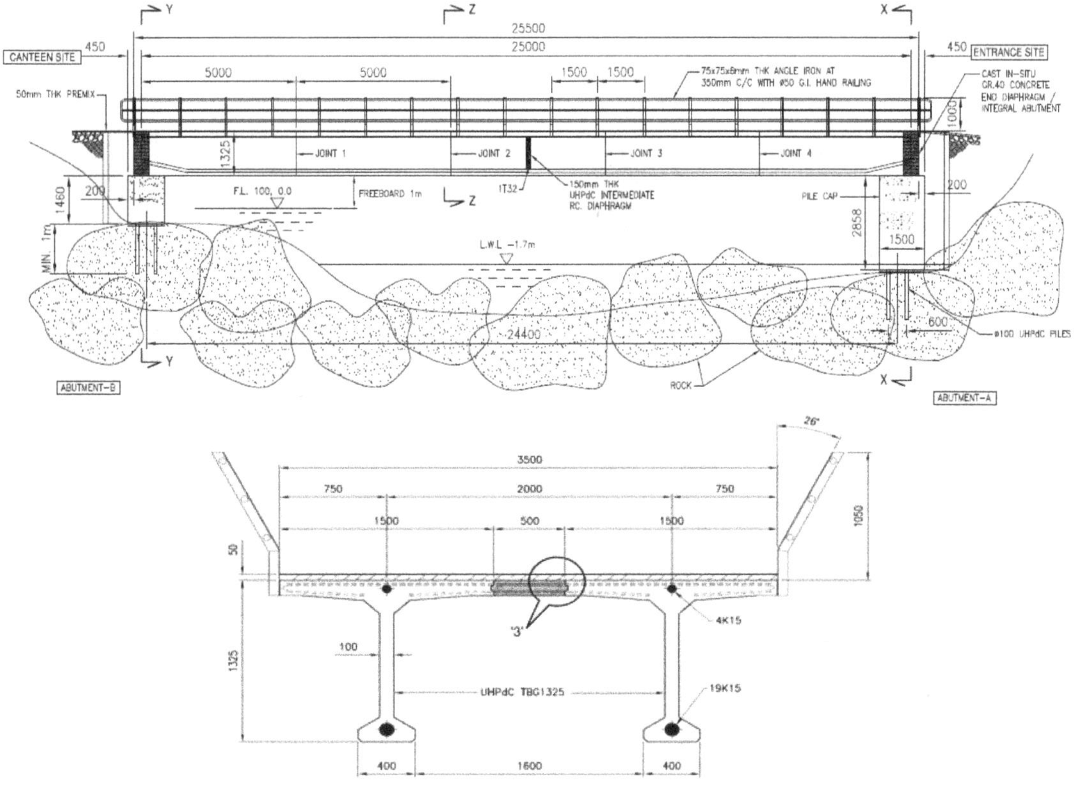

Figure 23: Typical sectional details of UHPFRC bridge at Kampung Ulu Geroh, Gopeng, Perak.

Figure 24a shows two units 20 tonnes capacity of mobile cranes were used to launch the girder. According to the beam launcher, they claimed this is the lightest concrete beam ever launched given the length of the girder is 25 m compared to the other beams. Figure 24b shows the bottom view of the two girders parked adjacent to each other with the joint ready to be stitched. UHPFRC bridge system is unique compare to other bridge system as the major part of the bridge deck was integrally casted together with the beam during manufacturing. Therefore, only small portion site required stitching work is required using the same grade of UHPFRC, after the beams have securely seated on the abutments. Figure 24c shows the in-situ UHPFRC was poured at the jointing area without any external compacting tools. After 1 day, the formwork was removed and the in-situ UHPFRC has attained an average cube compressive strength of 70 MPa (refer to Figure 24d). After 14 days, the cube sample of the in-situ stitch where tested to have an average cube compressive strength of 145 MPa.

Figure 24: (a) Beams launching using two units 20 tonnes capacity mobile crane, (b) in-situ bridge joint ready for stitching, (c) placing of in-situ UHPFRC for the bridge joints, (d) view from bottom of the bridge after stitching of the bridge joint and (e) the completed Kampung Ulu Geroh Bridge.

2.7   18 M KAMPUNG ULU KAMPAR BRIDGE (MEDIUM TRAFFIC BRIDGE), COMPLETED

Figure **25** shows another example on a short span bridge crossing a river with a total span length of 18 m was constructed by JKR Perak (Kinta Daerah). The bridge is located at a small village call Kampung Ulu Kampar, which is approximately 30 km from the capital city of Perak, Ipoh. Similar to the Kampung Ulu Geroh Bridge (refer to Section 2.6), this bridge also uses the full UHPFRC bridge/deck system. This bridge was designed to withstand 30 units HB loading and HA + KEL loading as per BD37/01. Construction of the bridge commenced at mid January 2012, and the bridge work was complete at end of February.

Similarly, the major challenge of the project is the poor access road to the job site. No long trailer is able to access to the job site. Although UHPFRC girder system has weight advantage over conventional system, the bridge designer or contractors still have bridge length issue to consider. The bridge designer eventually comes up with an idea to break the 18 m girder into three segments, thus having each segment measured 6 m long and weighted merely 6 tonnes. Thus a simple ten wheels truck can be used to transport the bridge segment (one at a time) and later on aligned off-site, then post-tensioned to form a single girder (see Figure 26).

Figure 25: The new UHPFRC bridge at Kampung Ulu Kampar, Gopeng, Perak.

(a)   (b)

Figure 26: (a) A 10 wheels truck transporting one 6 m long UHPFRC girder segment and (b) bridge segments aligned off-site ready for post-tensioning work.

2.8   51 M RANTAU-SILIAU BRIDGE (RANTAU, NEGERI SEMBILAN), ON-GOING

After the construction of Kampung Linsum bridge (refer to Section 2.5), the JKR Negeri Sembilan has decided to replace a multi-span old concrete bridge which span approximately 50 m. Figure 27a shows a recent photo of the existing bridge which has four rows of central RC columns (i.e. 5 columns per row) located at the waterway of the river. Figure 27b shows during the monsoon season, very often, large amount of debris trapped at the piers, which may not be an ideal practice due to large timber or logs may flow from the upstream and collide with the columns, thus eventually reduce or damage the structural integrity and safety of the bridge. JKR Negeri Sembilan took the full advantage of the UHPFRC technology and putting up a new single span 51 m long motorway bridge which comes with four carriageway lanes using five pieces of the same UHPFRC U-trough girder as presented in

Figure 21. The UHPFRC girder ends were encased in normal strength concrete abutments at the bridge site and made integral with the abutment seating. The bridge will be constructed without any central columns thus leaving the entire waterway of the river clear from obstruction. Detail of the new bridge is given in Figure 28. The bridge was designed to withstand full highway loading as per BD37/01.

(a)   (b)

Figure 27: (a) Existing old RC bridge with four rows of RC piers at the waterway of the river and (b) debris collected at the piers.

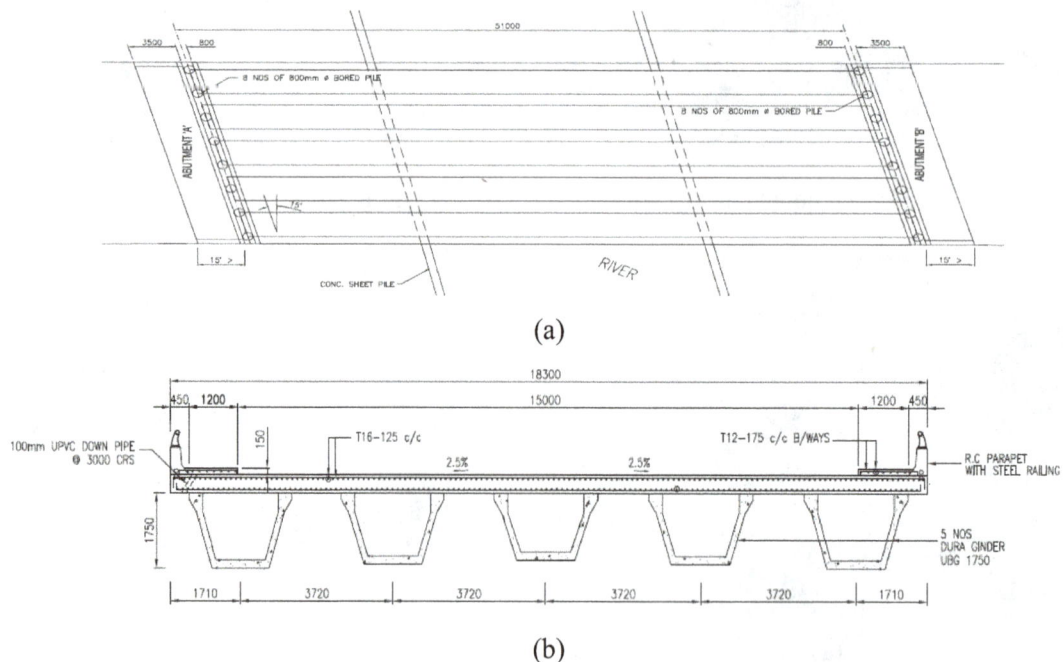

(a)

(b)

Figure 28: (a) Plan View and (b) Typical Bridge Section of new Rantau-Siliau Bridge.

2.9   50 M TITI BRIDGE (JELEBU, NEGERI SEMBILAN), ON-GOING

Figure 29 presents the general detail of a new dual carriageway motorway bridge that is currently under construction by JKR Negeri Sembilan. The composite bridge has a clear span of 50 m (leaving the entire waterway of the river clear from obstruction), and total bridge deck width of 11.9 m. The bridge was designed to withstand full highway loading as per BD37/01. Figure 29 shows the typical section of the bridge, whilst the bridge were constructed using three UHPFRC U-trough girders 1.75 m deep, 2.5 m wide at the top, topped with a 200 mm thick cast in-situ reinforced concrete. The UHPFRC girder ends was encased in normal strength concrete abutments at the bridge site and made integral with the abutment seating. The UHPFRC girder was built without any conventional shear reinforcements. Construction of the bridge was expected to complete by early 2013.

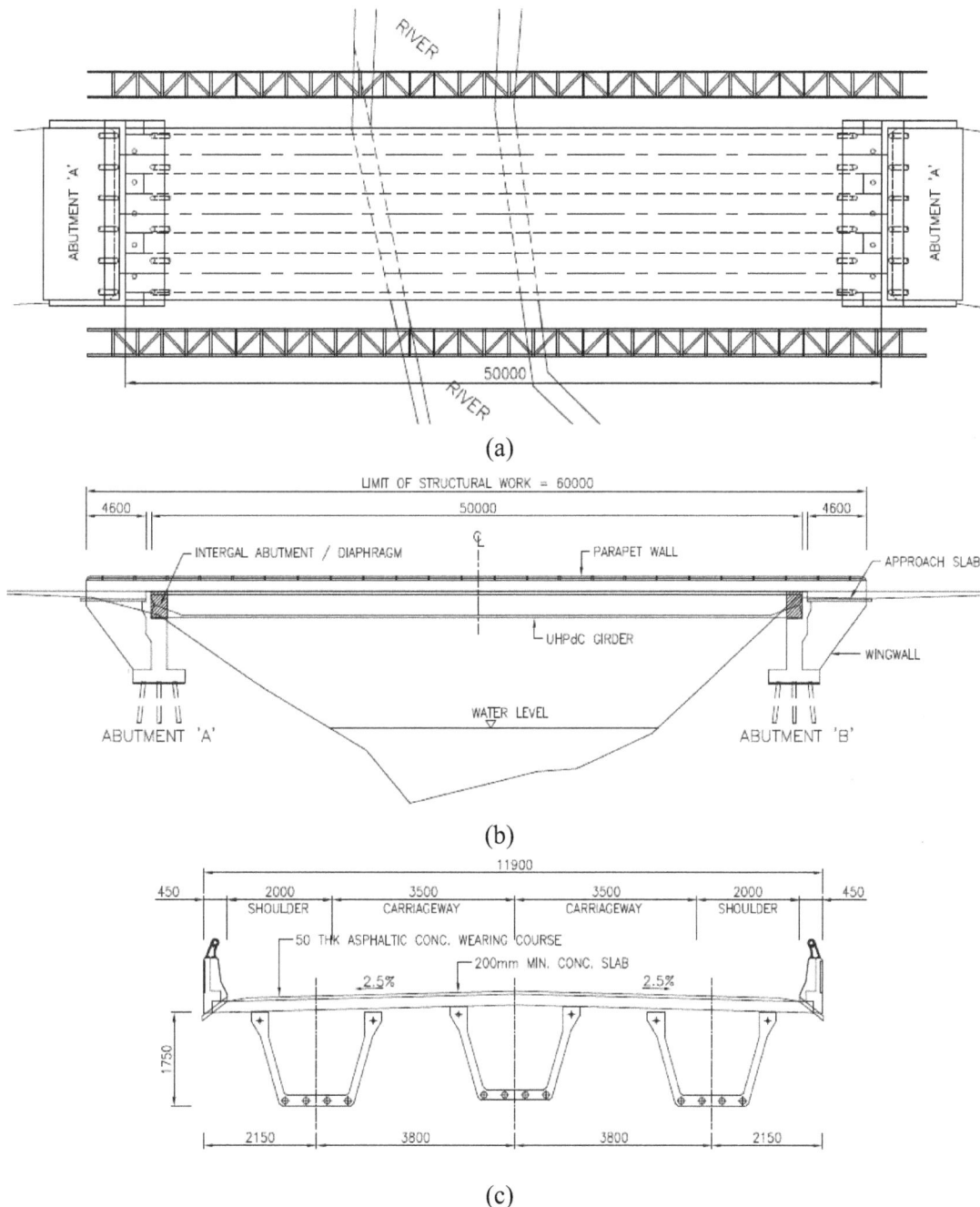

(a)

(b)

(c)

Figure 29: (a) Plan View, (b) Elevation View and (c) Typical Bridge Section of new Titi Bridge.

2.10   90 M SUNGAI NEROK BRIDGE (LENGGONG, PERAK), ON-GOING

To-date, the JKR Perak is *possibly* the world's first organization to use UHPFRC bridge girder in the construction of a multi-spans motorway bridge. Figure 30 shows the schematic detail of a newly awarded dual carriageway motorway bridge, which is currently under construction at Jalan Lenggong, crossing a river call Sungai Nerok (Kota Tampan Air). The bridge consists of three equal spans where each span has a span length of 30m c/c.

Similar to Kampung Ulu Geroh Bridge (refer to Section 2.6) and Kampung Ulu Kampar Bridge (refer to Section 2.7), this bridge is using the UHFRC bridge/deck system where the superstructure of the bridge is constructed without conventional RC deck. Figure 30 shows a total of 30 m UHPFRC Tee-girders used to construct the bridge. All the UHPFRC girders will be encased with normal concrete at the abutments/piers making the whole bridge as a full integral bridge without any expansion joint. This bridge was designed to withstand 45 units HB loading and HA + KEL loading as per BD37/01. Construction of the bridge is expected to commence at March 2012, and the bridge work is expected to complete before 2013.

According to the bridge designer, the superstructure of this bridge is approximately half the weight of the conventional design, which leads to significant saving in term of foundation. Besides that, the highly durable nature of UHPFRC promises to offer much longer design life and offers almost negligible maintenance during the service life of the bridge.

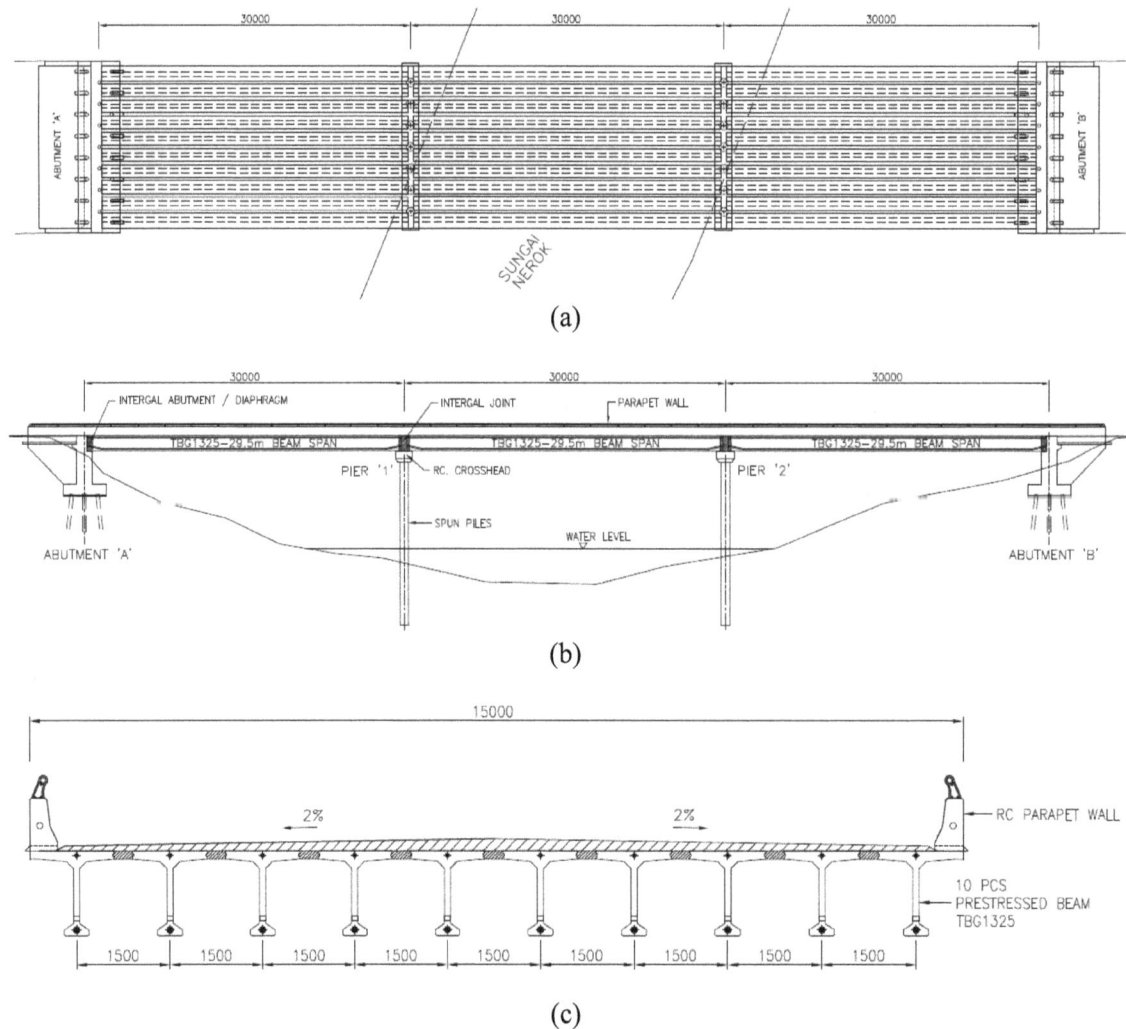

(a)

(b)

(c)

Figure 30: (a) Plan View, (b) Elevation View and (c) Typical Bridge Section of new Sungai Nerok Bridge, Lenggong, Perak.

2.11 29 M KAMPUNG BANIR BRIDGE (BATANG PADANG, PERAK), ON-GOING

Recently, the JKR Perak has called for tender on a dual carriageway bridge. Figure 31 shows the schematic detail of a newly tendered motorway bridge, which is located at Kampung Banir at Batang Padang. The bridge consists of single span crossing of 29 m. Similar to Kg. Ulu Geroh Bridge (refer to Section 2.6), Kg. Ulu Kampar Bridge (refer to Section 2.7) and Sungai Nerok Bridge (refer to Section 2.10), this bridge will be constructed using the UHFRC bridge/deck system where the superstructure of the bridge is constructed without conventional RC deck. All the UHPFRC girders will be encased with normal concrete at the abutments making the whole bridge as a full integral bridge without any expansion joint. This bridge was designed to withstand 45 units HB loading and HA + KEL loading as per BD37/01. Construction of the bridge is expected to commence at June 2012, and the bridge work is expected to complete before 2013.

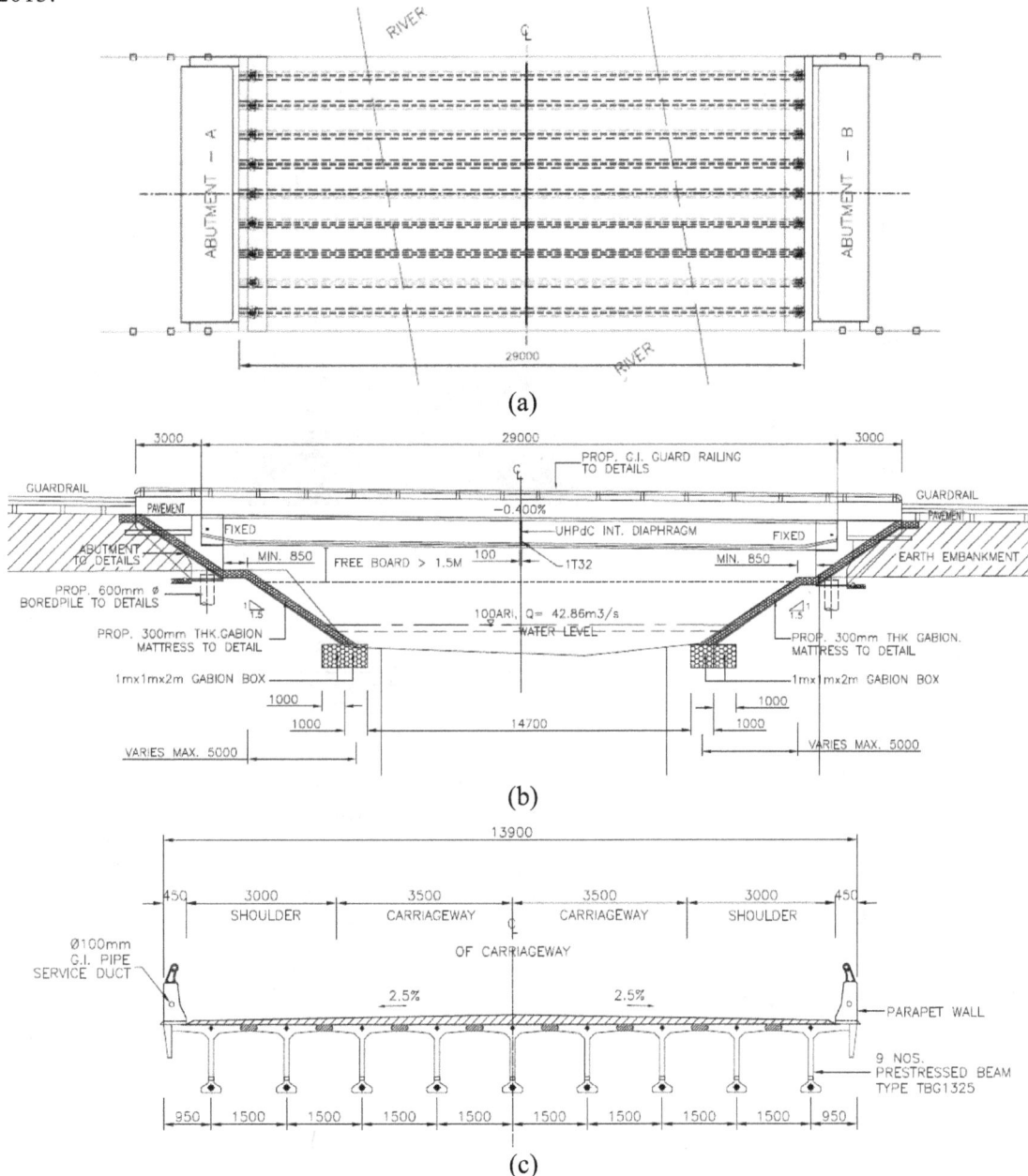

(a)

(b)

(c)

Figure 31: (a) Plan View, (b) Elevation View and (c) Typical Bridge Section of new bridge at Kampung Banir, Badang Padang, Perak.

2.12 173 M ARCHES FOR TOLL CANOPY (PULAI PENANG), ON-GOING

Recently, RMS Architect has designed a 150 m long by 30 m wide elliptical shape toll canopy at the Plus toll plaza of the Second Penang Bridge. The roof canopy is designed to be over-hanged by a pair of structural arches, come with arch length of 186 m, arch horizontal length of 173m and vertical height of 30 m (see

Figure 32). In this project, the project owner and concessionaries are the Lembaga Lebuhraya Malaysia (LLM) and Jambatan Kedua S/B, respectively. Initially the structural members of the arches were proposed to be built using conventional steel truss structure. However, it was later proposed with an alternative design where UHPFRC arch-like pipe structures were used because the owners have foreseen the benefit of using UHPFRC arch which includes mainly the elimination of maintenance as the arch structure is built close to the sea-side. In addition it gives immediate cost saving and improved aesthetics.

Figure 32: Artist impression of the proposed toll canopy and the arches.

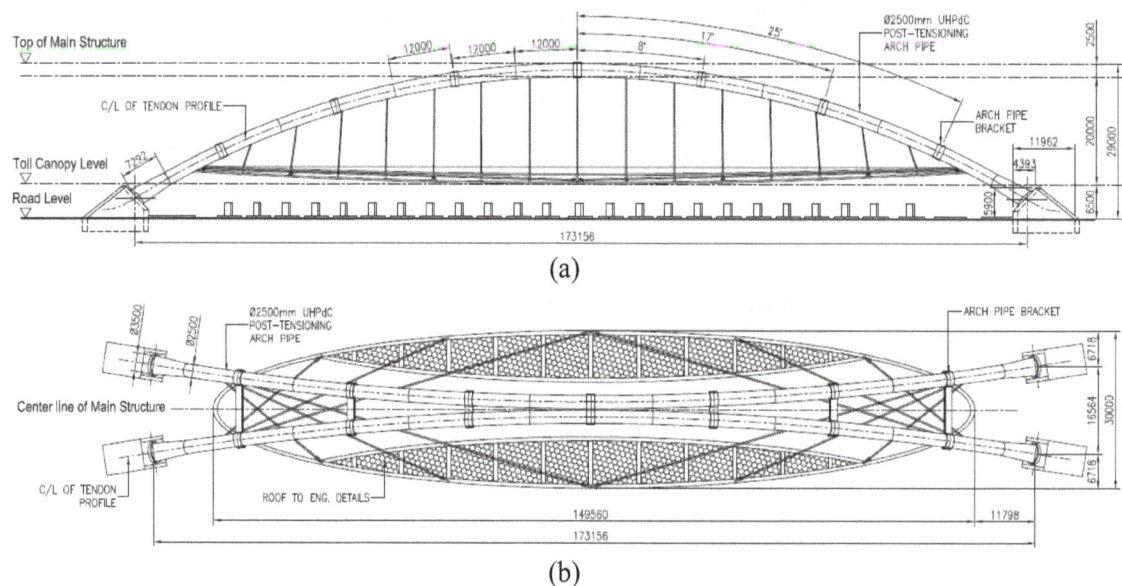

(a)

(b)

Figure 33: (a) Elevation View and (b) Plan View of proposed toll canopy.

3.0 CONCLUSION

This paper briefly presents some of the successful and on-going applications of UHPFRC technology in Malaysia. Some of the examples presented were the portal frame building, retaining wall, motorway bridges, security wall panel and arch structures. Throughout the examples presented herein, the technology of UHPFRC has proven to be an alternative sustainable construction material that embraced the uniqueness of both concrete and steel. Besides, the environmental impact assessment shows that UHPFRC structures are able to give immediate savings in terms of primary material consumption, embodied energy, CO_2 emissions and global warming potential. The UHPFRC technology is proved to be a greener construction material as it supports the vision of a sustainable construction in future. The authors are of the opinion that in the future, UHPFRC technology will contribute significantly to the realization of sustainable development. The technology carries an equation that sums up 'sustainable construction' in that it provides for a minimum impact on the environment, maximizes structural performance and provides a minimum total life-cycle cost solution. The benefits are:

- immediate reduction in overall consumption of non-renewable raw material;
- encourage the use of recycle materials (such as silica fume and GGBS);
- better quality and finishes of finishing products;
- prolong the service and design life of structures;
- minimized maintenance due to the its superior durability;
- support the visionary of green economy.

REFERENCES

[1] BS1881-Part 114, 1983. Testing concrete. Methods for determination of density of hardened concrete. *British Standard*, British Standards Institution, ISBN. 0-580-12948-9, 8 pp.

[2] AS 1012.9, 1999. Determination of the compressive strength of concrete specimens. *Australian Standard*, Standards Australia, 12 pp.

[3] BS6319-2, 1983. Testing of resin and polymer/cement compositions for use in construction. Method for measurement of compressive strength. *British Standard*, British Standards Institution, 4 pp.

[4] AS1012.16, 1996. Determination of creep of concrete cylinders in compression. *Australian Standard*, Standards Australia, 8 pp.

[5] BS1881-Part 121, 1983. Testing concrete. Methods for determination of static modulus of elasticity in compression. *British Standard*, British Standards Institution, 7 pp.

[6] BS EN 12390-6, 2000. Testing hardened concrete. Tensile splitting strength of test specimens. *British Standard*, British Standards Institution, ISBN: 0-580-36606-5, 14pp .

[7] ASTM C496, 2004. Standard test method for splitting tensile strength of cylindrical concrete specimens. *ASTM Standards*, ASTM International, United States, 5 pp.

[8] ASTM-C1018, 1997. Standard test method for flexural toughness and first crack strength of fiber reinforced concrete (using beam with third point loading). *ASTM Standards*, ASTM International, United States, 8 pp.

[9] ASTM C1202, 2005. Standard test method for electrical indication of concrete's ability to resist chloride ion penetration. *ASTM Standards*, ASTM International, West Conshohocken, PA, 6 pp.

[10] ASTM C1556, 2004. Standard test method for determining the apparent chloride diffusion coefficient of cementitious mixtures by bulk diffusion. *ASTM Standards*, ASTM International, United States, 7 pp.

[11] BS EN 14630, 2006. Products and systems for the protection and repair of concrete structures. Test methods. Determination of carbonation depth in hardened concrete by the phenolphthalein method. *British Standard*, British Standards Institution, ISBN: 0-580-49622-8, 12 pp.

[12] ASTM C944-99, 2005. Standard test method for abrasion resistance of concrete or mortar surfaces by the rotating-cutter method. *ASTM Standards*, ASTM International, United States, 4 pp.

[13] BS1881-Part 122, 1983. Testing concrete. Method for determination of water absorption. *British Standard*, British Standards Institution, ISBN: 0-580-12959-4, 4 pp.

[14] Adeline, R., Iachemi, M., & Blais, P. (1998). Design and behavior of the Sherbrooke Footbridge. Paper presented at the the International Symposium on High-Performance and Reactive Powder Concretes, Sherbrooke, Canada.

[15] Deem, S. (2001). Concrete Attraction-Something New on the French Menu-Concrete. Retrieved 10 June 2009, from http://www.popularmechanics.com/science/research/2002/6/concrete/print

[16] Cavill, B., & Chirgwin, G. (2003). The world's First RPC Road Bridge at Shepherds Gully Creek, NSW. Paper presented at the 21st Biennial Conference of the Concrete Institute of Australia (CIA), NSW, Australia.

[17] Graybeal, B. A. (2011). FHWA TECHNOTE: Ultra High Performance Concrete, FHWA Publication No: FHWA-HRT-11-038: Federal Highway Administration.

[18] Voort, T. L. V., Suleiman, M. T., Sritharan, S., Iowa Highway Research Board, & Iowa Dept. of Transportation. (2008). Design and Performance Verification of UHPC Piles for Deep Foundations: Center for Transportation Research and Education, Iowa State University.

[19] Massicotte, B., & Boucher-Proulx, G. (2010). Seismic Retrofitting of Bridge Piers with UHPFRC Jackets Designing and Building with UHPFRC: State of the Art and Development (pp. 531-540). London: Wiley-ISTE.

[20] Brühwiler, E., & Denarié, E. (2008). Rehabiltation of Concrete Structures Using Ultra- High Performance Fiber-Reinforced Concrete Proceedings of the Second International Symposium on Ultra-High Performance Concrete. Kassel, Germany.

[21] Schmidt, C., Riedl, S., Geisenhanslüke, C., & Schmidt, M. (2008). Strengthening and Rehabilitation of Pavements Applying Thin Layers of Reinforced Ultra-High Performance Concrete (UHPC-White Topping) Proceedings of the Second International Symposium on Ultra-High Performance Concrete. Kassel, Germany.

[22] Green, B. (2010). An Investigation of UHPC/RPC Materials for Enhanced Penetration Resistance Paper presented at the the American Concrete Institute Fall Convention.

[23] Rebentrost, M., & and Wight, G. (2009). Investigation of UHPFRC Slabs Under Blast Loads. Paper presented at the Proceedings, Ultra-High Performance Fiber Reinforced Concrete Marseille, France.

[24] Voo, Y. L., & Foster, S. J. (2009). Reactive powder concrete: analysis and design of RPC girders. Germany: Lambert Academic Publishing.

[25] Voo, Y. L., and Poon, W.K., (2008), "The World First Portal Frame Building (Wilson Hall) Constructed Using Ultra-High Performance Concrete", 33rd Conference on "Our World in Concrete & Structures: Sustainability", 25 -27 August 2008, Singapore, pages: 493 –500.

[26] Poon, W.K., Voo, Y.L., and Yap, S.K, (2009), "Thin Protective Wall Panel Made From 150 MPa Ultra-High Performance Ductile Concrete (UHPdC-DURA)" Proceedings of the 10th International Conference on Concrete Engineering and Technology 2009 (CONCET 2009), 02-04 March, Shah Alam, Kuala Lumpur, Malaysia. (in CD-format).

[27] Voo, Y.L., Augustin, P.C., and Thamboe, T.A.J., (2011), "Construction and Design of a 50 m Single Span UHP Ductile Concrete Composite Road Bridge", The Structural Engineer, Vol 89(15), 2nd August, The Institution of Structural Engineers, UK, pages: 24–31

FLEXURAL BEHAVIOR OF STEEL-FIBER-ADDED-RC (SFARC) BEAMS WITH C30 AND C50 CLASSES OF CONCRETE

Hamid Pesaran Behbahani[1], Behzad Nematollahi*[2], Abdul Rahman Mohd. Sam[3], F. C. Lai[4]

[1]Faculty of Civil Engineering, Universiti Teknologi Malaysia (UTM), Skudai, Malaysia
[2]Civil Engineering Department, Universiti Putra Malaysia (UPM), Serdang, Malaysia
[3]Faculty of Civil Engineering, Universiti Teknologi Malaysia (UTM), Skudai, Malaysia
[4]Regional Technology Support Centre, Sika Kimia Sdn Bhd, Malaysia

*Corresponding E-mail : behcom62@gmail.com

ABSTRACT

Although conventional reinforced concrete (RC) is the most globally used building material; however, its detrimental structural characteristics such as brittle failure mechanism in tension need to be improved. Discrete and short steel fibers (SFs) can be added into the concrete mix to improve its brittleness. The effects of the addition of optimum percentage of SFs on flexural behavior of RC beams have been investigated in this paper. In this study, the optimum percentage of hooked-end SFs with the dimensions of 0.75 mm in diameter and 50 mm in length are added in RC beams with two different classes of concrete (i.e. two different compressive strengths of 30MPa (C30) and 50MPa (C50)). In order to determine the optimum percentage of SFs added to the concrete mix, 15 prisms and 30 cubes with 5 different percentages of SFs (i.e. 0%v/v, 0.5%v/v, 1%v/v, 1.5%v/v, and 2%v/v) from both C30 and C50 classes of concrete have been tested. Based on the results of the flexural strength and compressive strength tests, it is found that the optimum value is 1% by volume (i.e. 78.5 kg/m^3) for the specific type of fiber used in this study. Subsequently, to investigate the flexural behavior of steel fiber added RC (SFARC) beams compared to conventional RC beams with no SFs, two RC beams with the dimensions of 170 mm in height, 120 mm in width, and 2400 mm in length, with the SF percentages of 0 and 1%v/v and both having exactly the same steel reinforcement were tested under flexure using a four-point loading test setup for both C30 and C50 classes of concrete. The experimental results show that the SFARC beams with 1% by volume of the SFs have higher first cracking strength, ultimate flexural strength, stiffness, and ductility compared to that of the conventional RC beams with no SFs. Furthermore, the addition of the SFs has more effects on the RC beams with higher compressive strength (50 MPa) compared to lower concrete grade (30 MPa).

Keywords: *Optimum steel fiber percentage, Steel fiber added reinforced concrete (SFARC) beam, Flexural behavior*

1.0   INTRODUCTION

Structural concrete is by far one of the most extensively used construction material all over the world. Concrete is a brittle material, with little capacity to resist tensile stresses/strains without cracking. Thus it requires reinforcement before it can be used as construction material. Historically, this reinforcement has been in the form of continuous reinforcing bars which could be placed in the structure at the appropriate locations to withstand the imposed tensile and shear stresses. Fibers, on the other hand, are generally short, discontinuous, and randomly distributed throughout the concrete to produce a composite construction material known as fiber reinforced concrete (FRC). Steel fiber (SF) is the most popular type of fiber used in reinforced concrete.

Fibers are used to prevent/control plastic and drying shrinkage in concrete and significantly increase its flexural toughness, the energy absorption capacity, ductile behaviour prior to the ultimate failure, reduces cracking, and improves durability [1]. Steel fiber reinforced concrete

(SFRC) also known as steel-fiber-added concrete (SFAC) has extensive applications such as tunnel linings, large slabs and floors with great live load, rock slope stabilization, dam constructions, composite metal decks, seismic retrofitting, repairs and rehabilitations of marine structures, fire protection coatings, concrete pipes as well as conventional RC frames due to its superior toughness against dynamic loads compared to conventional reinforced concrete [1, 2]. The objectives of this study are as follows:

1) To determine the optimum percentage of the specific type of SFs used in this study for C30 and C50 classes of concrete based on the cube compressive strength and flexural strength tests with five different SF dosages of 0%v/v, 0.5%v/v, 1%v/v, 1.5%v/v, and 2%v/v.

2) To determine the structural flexural behaviour of steel-fiber-added reinforced concrete (SFARC) beams of C30 and C50 classes of concrete, containing the optimum percentage of the SFs as determined in the previous step, compared against the conventional RC beams of the same concrete grades.

2.0   BACKGROUND OF STUDY

Hannant [3] found that the addition of steel fibers has more influence on the flexural strength of concrete compared to its tensile/compressive strength. As reported by Oh et al. [4], the flexural strength of SFRC has been increased about 55% with the addition of 2% by volume of steel fibers. Johnston [5, 6] has found that the compressive strength of SFRC is increased about 20% with the addition of 1.2% by volume of steel fibers. Research conducted by Johnston [7] showed that the compressive strength of SFRC has been increased from 0 to 15% with the addition of up to 1.5% of steel fibers by volume. In a research conducted by Hartman [8] twelve different SFRC beams containing two different SFs amount of 60 and 100 kg/m^3 of Dramix RC-65/35-BN type were tested and it was concluded that the ratio of the experimental ultimate load to the theoritical ultimate load was bigger for those SFRC beams having a 60 kg/m^3 amount of SFs. Narayanan and Darwish [9] reported that the mode of failure has changed from shear to flexure when the percentage of steel fibers was increased beyond 1.0%.

According to the literature, it can be understood that the optimum percentage of SFs in SFARC beams should be within the range of 1 to 2.5% by concrete volume and depends on the type of the fiber (i.e its aspect ratio) and properties of the fresh concrete such as workability and its maximum size aggregates. The addition of less than 1% by voulme of SFs is inadequate and amounts beyond 2.5% is also ineffective mostly due to the inadequate compaction and localized distribution of the fibers in the concrete mix leading to a considerable reduction in the compressive strength compared with the same grade of concrete [1].

To the best of our knowledge, there is only one extensive research regarding the optimum percentage of SF for SFARC beams and the economic study for the usage of SFs in conventional RC beams conducted by Altun et al. [1]. All the previous research is conducted with only one grade of concrete, and the effects of SFs on different classes of concrete and the type of fibers have not been considered. The flexural behaviour and toughness of conventional RC and SFARC beams for C20 and C30 classes of concrete with three different SF dosages of 0, 30, 60 kg/m^3 have been investigated in their study in a comparative way. It was concluded that both the ultimate loads and the flexural toughness of SFARC beams manufactured with concrete grades of 20 MPa and 30MPa with 30 kg/m^3 of SFs increased significantly compared to those conventional RC beams and the addition of 30 kg/m^3 of steel fibers is better than that of 60 kg/m^3 for the SFARC beams.

3.0   EXPERIMENTAL WORK

3.1.   MATERIALS AND MIX PROPORTIONING

Two concrete grades of 30 and 50MPa were used in this study. The mix design was based on the guidelines given in the Design of Normal Concrete Mixes [10] with slightly modification to adapt to the available local raw materials. To achieve concrete grade of 50 MPa water cement ratio has been reduced from 0.56 to 0.4 by using the superplasticizer. The concrete mix design used in this study is shown in Table 1. The steel fibers used in this study are of Wirand® Fiber-FF3 type with a diameter of 0.75 mm and a length of 50 mm as shown in Figure 1, and its strain at failure and tensile strength are less than 4% and beyond 1100 MPa, respectively.

To ensure that the SFs have been distributed uniformly in the concrete matrix, the rate of SFs addition into the concrete mixer was 20 kg/min, and then the mixer has rotated at high speed for 5 minutes, as recommended by the relevant RILEM publications [11, 12].

Table 1: Mix Proportioning of the Concrete with Grades of 30 and 50

Ingridients	Mass (kg/m^3)	
	Grade 30 MPa	Grade 50 MPa
W/C Ratio	0.56	0.4
Mixed Water	230	164
Portland Cement	410	410
Fine Aggregate (Sand)	902	780
Coarese Aggregate (Max size = 16 mm)	833	1040
Superplasticizer	0	1.4 Liter/100 kg of cement

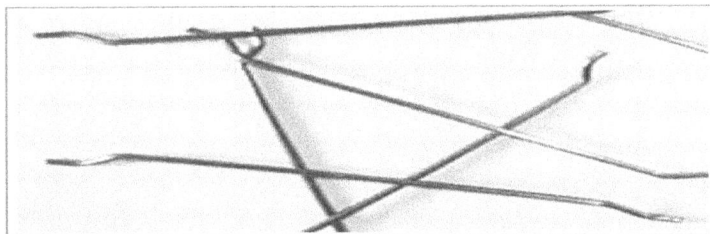

Figure 1: Hooked-end shape steel fibers used in this Study

3.2.   OPTIMUM PERCENTAGE OF SFS IN C30 AND C50 CLASSES OF CONCRETE

To determine the optimum percentage of SF for C30 and C50 classes of concrete, cube compressive strength and flexural strength tests with five different SF dosages of (0%, 0.5%, 1%, 1.5%, 2%)v/v were carried out. Vebe time test was carried out to determine workability of the mix designs.

3.2.1. CUBE COMPRESSIVE STRENGTH TEST

The compressive strength of the specimens at the age of 28 days was measured in a test setup in accordance with BS 1881: Part 116 [13]. For each concrete grade of 30 MPa and 50 MPa, thirty cubic specimens with the dimensions of 150 × 150 × 150 mm were prepared. Six cubic specimens for each of the 5 different SF dosages of (0%, 0.5%, 1.0%, 1.5%, and 2%) v/v were cast. A hand poker vibrator was used for compaction of concrete in cubic specimens. All 60 cubes were de-moulded after one day and immersed in a water tank for a period of 28 days to assure adequate curing.

3.2.2. FLEXURAL STRENGTH TEST

The flexural strength test was conducted in accordance with ASTM C1018-97 [14]. For each of the different SF dosages of (0%, 0.5%, 1.0%, 1.5%, and 2%)v/v, three prisms with the dimensions of 150 × 150 × 750 mm for each of the C30 and C50 classes of concrete were prepared. A hand poker vibrator was used for compaction of concrete in prisms. All 30 prisms were de-moulded after one day and immersed in the water tank for a period of 28 days to assure adequate curing. After 28 days, each prism was tested under flexure using the four point loading test setup. The specifications of the cubes and prisms used to determine the optimum percentage of the SFs in C30 and C50 classes of concrete are summarized in Table 2.

Table 2: characteristics of the cubic and prism specimens used in this study

Volumetric ratio of the SF	Classes of concrete	Number of specimens	
		Cubes (150 ×150 × 150 mm)	Prisms (150 × 150 × 750 mm)
0	C30-0.0%	6	3
	C50-0.0%	6	3
0.5	C30-0.5%	6	3
	C50-0.5%	6	3
1	C30-1.0%	6	3
	C50-1.0%	6	3
1.5	C30-1.5%	6	3
	C50-1.5%	6	3
2	C30-2.0%	6	3
	C50-2.0%	6	3

3.2.3. VEBE TIME TEST

Vebe time test was used to measure the workability of the concrete mix in this study. The Vebe time test determines the remoulding ability of concrete under vibration. The test results reflect the time required to remould an amount of concrete under given vibration situation.

3.3. STRUCTURAL FLEXURAL BEHAVIOR OF SFARC BEAMS

For each concrete grade of 30 and 50 MPa, two beams with the dimensions of 120 × 170 × 2400 mm were cast with the optimum percentage of SFs, as determined in the previous tests, added to one of the beams of each concrete grade to prepare the SFARC beams with two different classes of concrete. The other beam of each grade was cast as conventional RC beams with no SFs. The dimensions, concrete mix, and the steel reinforcement used in both beams of each grade were all exactly the same and the only difference was the addition of the optimum SF dosage in one of the beams. The beam's reinforcement inside wooden moulds before casting is shown in Figure 2. Figure 3 shows the cross section of the RC and the SFARC beams used in this study. As shown in this figure, two steel bars with the diameter of 12 mm and characteristic yield strength of 460 N/mm^2 were used in bottom of the section and two hanger bars with the diameter of 6 mm and characteristic yield strength of 250 N/mm^2 were used at the top. The shear links with a diameter of 6 mm and characteristic yield strength of 250 N/mm^2 were arranged at 100 mm centre to centre. The specifications of the RC and the SFARC beams tested are summarized in Table 3. Proper supports for the wooden moulds were provided to prevent mould deformation during vibration and compaction process. All the four beams were de-moulded after one day and kept completely covered by burlap. Regularly all the beams were splashed with lots of water while covered up under burlap for a period of 28 days. All the four beams were tested at the age of 28 days in a four-point loading test setup as shown in Figure 4. As shown in this figure, the beam was positioned as simply supported. The load was transmitted to the beam through a hydraulic jack placed at the centre of the spreader beam and measured by the load cell located between the spreader beam and the hydraulic jack. Three linear variable differential transducers (LVDTs) were installed using a magnetic stand at the soffit of the beam in the mid-span as well as the two points where the loads were acted. The LVDTs and the load cell were connected to a data logger. The beams were loaded to failure. The rate of load increase was 0.5 kN (i.e. for every 0.5 kN increase in load, corresponding deflection were measured through the LVDTs. Figure 5 shows a picture taken from the test setup used in this study.

Figure 2: Appearance of the RC and the SFARC beams before casting

Table 3: Specifications of the RC and the SFARC beams used in this study

Beam type	No. of beam specimens (120 × 170 × 2400 mm)	
	C30	C50
RC	1	1
SFARC with Opt. percentage of SFs	1	1

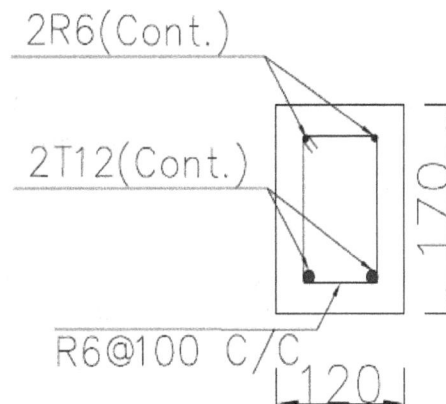

Figure 3: Cross Section of the RC and the SFARC Beams (dimensions in mm)

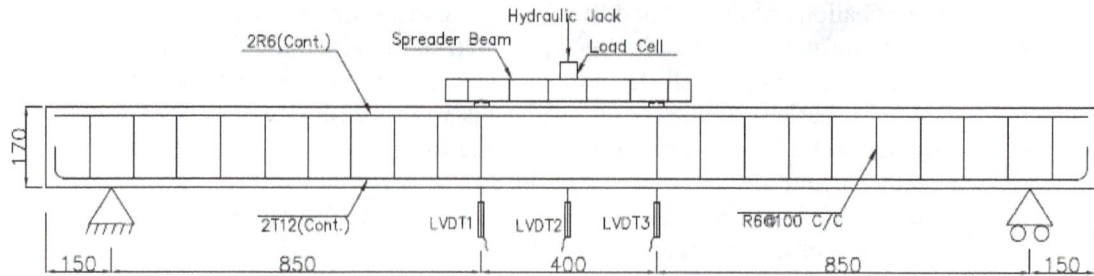

Figure 4: Schematic four-point loading test setup used in this study (dimensions in mm)

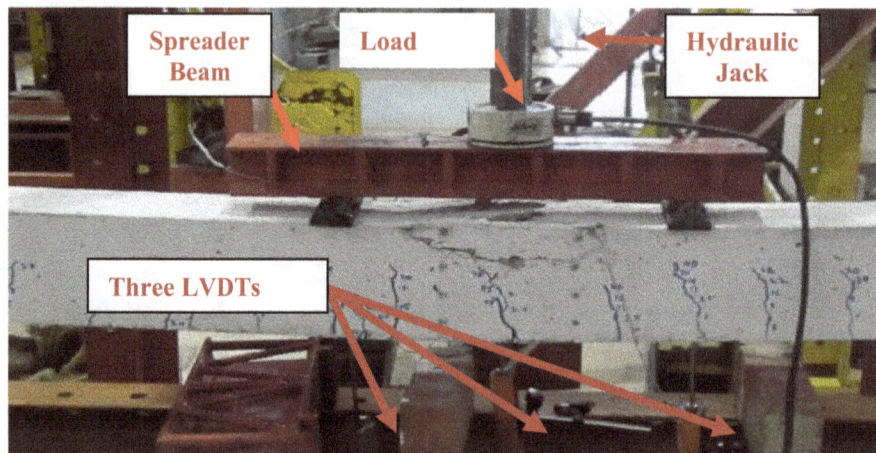

Figure 5: Four-point loading test setup used in this study

4.0   RESULTS AND DISCUSSION

4.1.   OPTIMUM PERCENTAGE OF SF IN C30 AND C50 CLASSES OF CONCRETE

The mean values of cube compressive strength of the specimens at the age of 28 days are shown in Table 4. Meanwhile, the results of the four-point loading test on the prisms of C30 and C50 classes of concrete are shown in Tables 5 and 6, respectively. Table 7 shows the results of the Vebe time test.

Table 4: Mean values of cube compressive strength of specimens at the age of 28 days

Specimen label	Mean value of cube compressive strength at the age of 28 days (MPa)	
	C30	C50
RC	32.3	54.8
SFARC-0.5%	37.5	66.3
SFARC-1.0%	38.2	68.5
SFARC-1.5%	35.4	63.0
SFARC-2.0%	32.5	55.1

Table 5: Mean values of flexural strength of prisms of C30 class of concrete at the age of 28 days

Concrete class		Flexural strength (MPa)	Mid-span deflection at ultimate load (mm)
C30	RC	18.0	0.73
	SFARC-0.5%	18.3	0.68
	SFARC-1.0%	19.9	1.12
	SFARC-1.5%	16.9	2.23
	SFARC-2.0%	16.7	2.78

Table 6: Mean values of flexural strength of prisms of C50 class of concrete at the age of 28 days

Concrete class		Flexural strength (MPa)	Mid-span deflection at ultimate load (mm)
C50	RC	18.5	0.81
	SFARC 0.5%	19.0	0.75
	SFARC 1.0%	22.0	1.30
	SFARC 1.5%	17.9	2.55
	SFARC 2.0%	17.5	2.95

As it is obvious in Tables 4, 5 and 6, in both C30 and C50 classes of concrete, specimens containing 1% by volume of the SFs have the highest flexural and compressive strength as compared to other specimens with different percentage of the SFs. In addition, as shown in Table 7, the addition of 1% of the SFs did not have a considerable effect on concrete workability compared to the other SF dosages. The increase in flexural strength can be due to crack-arrest and crack-control mechanism of concrete with the addition of SFs. Furthermore, as shown in Tables 4, 5 and 6 the addition of more than 1% by volume of SFs has led to a decrease in both cube compressive and flexural strength of concrete. This can be due to an appreciable drop in workability of concrete mix with the addition of more than 1% by volume of SFs causing physical difficulties during compaction and casting process of specimens. Therefore, it can be concluded that for this specific type of fibers used in this study the optimum addition of SFs in concrete mix is 1% by volume with respect to both cube compressive and flexural strength.

Table 7: Vebe time test of fresh concrete

Concrete Class	Measured Vebe time of fresh concrete (S)				
	0%	0.5%	1.0%	1.5%	2.0%
C30	1.5	1.75	3.4	7.5	9.8
C50	2.0	2.4	5.0	11.1	14.0

4.2. FLEXURAL BEHAVIOUR OF RC AND SFARC BEAMS

4.2.1. FLEXURAL FIRST CRACKING STRENGTH AND MODULUS OF RUPTURE OF RC AND SFARC BEAMS

The results of flexural test using the four-point loading setup on RC and SFARC beams with 1% of SF by volume are shown in Table 8. As shown in this table, the addition of 1% by volume of SFs (the optimum percentage of SFs used in this study as determined in previous step) led to an increase of 13% and 25% in flexural first cracking load and an increase of 7% and 15%

in ultimate experimental load of SFARC beams of C30 and C50 classes of concrete, respectively. Therefore, it can be concluded that the addition of 1% of SFs in RC beams leads to a higher increase in both flexural first cracking load and ultimate experimental load of beams with higher concrete grade (50 MPa) compared to lower grade of concrete (30 MPa). This can be due to the lower water cement ratio of higher concrete grade (50 MPa) causing less voids in concrete mix and an increase in the fiber-matrix bond strength.

Table 8: Flexural first cracking strength and modulus of rupture of RC and SFARC beams

Beam type	Flexural 1st cracking load (kN)		Ultimate experimental load (kN)	
	C30	C50	C30	C50
RC	7.5	8.0	41.0	42.0
SFARC-1.0%	8.5	10.0	44.0	48.4

4.2.2. LOAD VERSUS DEFLECTION OF RC AND SFARC BEAMS

Figure 5 shows the load versus mid-span displacement of the RC and SFARC beams. The area underneath the graph represents the fracture energy and ductility of concrete beam. As shown in Figure 5, the area underneath load versus deflection graphs of the SFARC beams is bigger than their counterpart RC beams. Thus, it can be concluded that the SFARC beams have more ductility compared to the conventional RC beams. The mid-span displacements at ultimate flexural load of the beams are shown in Table 9. As shown in this table, although the addition of 1% by volume of SFs led to an increase in the ultimate flexural load; however, SFARC beams have lesser deflection compared to their counterpart RC beams. It can be concluded that the stiffness of the SFARC beams is enhanced in comparison with the RC beams.

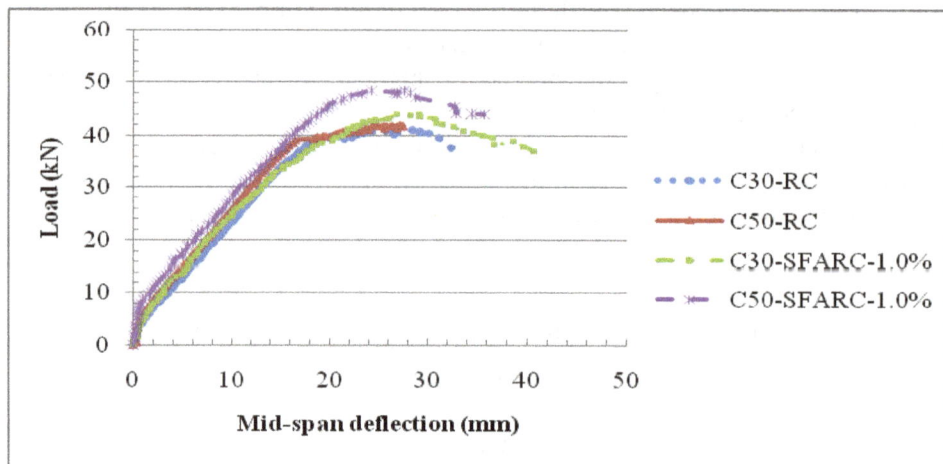

Figure 5: Load Vs. mid-span deflection graph of RC and SFARC beams

Table 9: Mid-span deflection at ultimate flexural load

Beam type	Ultimate flexural load (kN)		Mid-span displacement at ultimate load (mm)	
	C30	C50	C30	C50
RC	41	42	28.46	25.58
SFARC-1.0%	44	48.4	26.9	24.2

4.2.3. FAILURE MODE OF RC AND SFARC BEAMS

The failure mode of the RC and SFARC beams are shown in Figures 6. As it can be seen in these figures, both the RC and the SFARC-1.0% beams have failed in flexural-tension mode as predicted during the design stage. After failure, the number of cracks and the average crack width of each beam are registered in Table 10.

Table 10: Number of cracks and average crack width in RC and SFARC-1.0% beams

Beam type	Number of cracks at failure		Average crack width (mm)	
	C30	C50	C30	C50
RC	19	20	8.2	8.9
SFARC-1.0%	22	25	7.8	6.9

According to Table 10, the addition of 1.0% by volume of SF has led to an increase of 16% and 25% in number of cracks and a decrease in the average crack width of the SFARC beams of C30 and C50 classes of concrete, respectively. Therefore, it can be concluded that the addition of 1.0% by volume of steel fibers in RC beams increases the number of cracks and decreases the average crack width due to the higher ductility behaviour of SFARC beams. Furthermore, the addition of SF has more effect in SFARC beams with higher compressive strength (C50) compared against the lower concrete grade (C30).

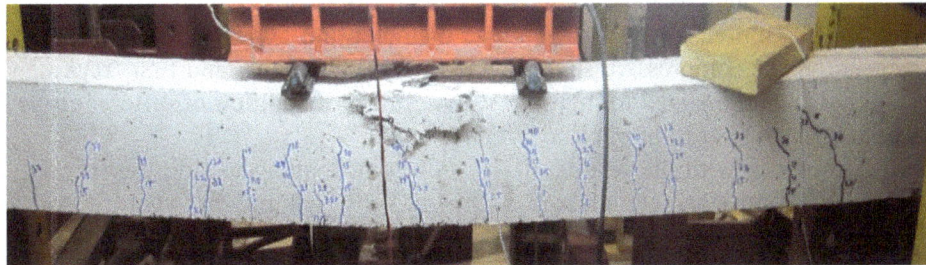

(a)

(b)

(c)

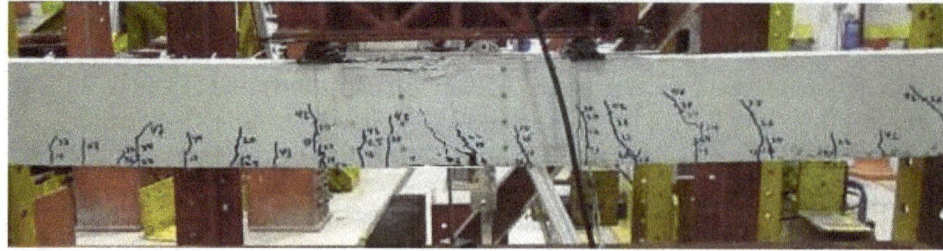

(d)

Figure 6: Failure mode and crack pattern of the RC and the SFARC beams; (a)-C30-RC beam, (b)- C50-RC beam, (c)-C30-SFARC-1.0% beam, (d)-C50-SFARC-1.0% beam

5.0 CONCLUSIONS

According to the results of the cube compressive strength test, it is observed that the cube compressive strength of specimens made from C30 and C50 classes of concrete with addition of 1.0% by volume of the SF has increased appreciably compared to other specimens with different percentage of steel fibers. Based on the results of flexural strength test, it is concluded that both the first cracking strength and flexural toughness of prisms made from C30 and C50 classes of concrete with addition of 1.0% by volume of the SFs has increased considerably as compared to those prisms with different percentage of the SFs. Therefore, it can be concluded that in SFARC beams made from concrete grade of 30 MPa and 50 MPa, the optimum percentage of the hooked-end SFs with the dimensions of 0.75 mm in diameter and 50 mm in length is 1.0% by volume with respect to cube compressive strength and flexural toughness and first cracking strength tests. Future study is recommended to investigate the effects of addition of different type of SFs into a high workable steel fiber reinforced concrete.

In accordance with the results of the four-point loading test on the SFARC beams made from C30 and C50 concrete classes with 1.0% by volume of steel fibers, the following conclusions can be drawn:

(i)   Addition of 1.0% by volume of SFs in the RC beams increases both the flexural first-cracking strength and flexural toughness of the SFARC beams and leads to an appreciable increase in their ductility and stiffness compared to those conventional RC beams without addition of SFs.

(ii)   Addition of 1.0% by volume of SFs has more effects on the RC beams made from the concrete with higher compressive strength (C50) compared to the concrete with lower compressive strength (C30) due to the better bonding between the fibers and the concrete paste.

(iii)   1.0% by volume of the SFs could be applied in the RC beams to get better flexural behaviour.

ACKNOWLEDGMENTS

The support of the Structures and Materials Laboratory, Faculty of Civil Engineering, Universiti Teknologi Malaysia (UTM) for their assistance in conducting the experimental work is gratefully acknowledged. This research was financed by Zamalah Institutional Scholarship provided by Universiti Teknologi Malaysia (UTM) and the Ministry of Higher Education of Malaysia. The authors are grateful to Dr. Yen Lei Voo, CEO and Director of Dura Technology Sdn. Bhd. for reading through the manuscript and for making several invaluable comments and suggestions to improve the quality of the article.

REFERENCES

[1] F. Altun et al. (2006) "Effects of steel fiber addition on mechanical properties of concrete and RC beams" Construction and Building Materials Vol 21, pages: 654–661.

[2] Ocean concrete products, Ocean Heidelberg Cement Group (1999). Steel Fibre Reinforcement – Working Together to Build Our Communities Report.

[3] Hannant, D.J. (1978).Fiber Cements and Fiber Concrete. John Wiley and Sons Ltd., Chichester, UK, page 53.

[4] Oh,S.G., Noguchi T. and Tomosawa,F. (1999). "Evaluation of Rheological Constants of High-Fluidity Concrete by Using the Thickness of Excess Paste." Journal of the Society of Materials Science, August, Japan.

[5] Johnston, C.D. (1982). "Definition and Measurement of Flexural Toughness Parameters for Fiber Reinforced Concrete" Cem. Concr. Agg.

[6] Johnston, C.D, Colin, D. (1982). "Fibre Reinforced Concrete" Progress in Concrete Technology CANMET, Energy, Mines and Resources, Canada, pages 215-236.

[7] Johnston, C.D. (1974). "Steel Fiber Reinforced Mortar and Concrete. A Review of Mechanical Properties, in fiber reinforced concrete" ACI – SP 44 – Detroit.

[8] Hartman, T. (1999). Steel Fiber Reinforced Concrete. Master Thesis., Department of Structural Engineering. Royal Institute of Technology, Stockholm, Sweden.

[9] Narayanan, R., and Darwish, I. (1987). "Bond Anchorage of Pretentioned FRP Tendon at Force Release."ASCE Journal of Structural Engineering,Vol 118 (10),pages 2837-2854.

[10] Design of Normal Concrete Mixes, Department of Environment (1988). United Kingdoms.

[11] RILEMTC162-TDF. Test and design methods for steel fibre reinforced concrete: re-design method. Mater Struct 2000, Vol 33 (March), pages 75–81.

[12] RILEM TC162-TDF. Test and design methods for steel fibre reinforced concrete: bending test. Mater Struct 2000, Vol 33 (January–February), pages 3–5.

[13] British Standard Institution (1983). Method for Determination of Compressive Strength of Concrete Cubes. London. BS 1881: Part 116. 73.

[14] ASTM C1018-97 (1997), standard test method for flexural toughness and first-crack strength of fiber-reinforced concrete (using beam with third-point loading), ASTM Standards.

4

DESIGN AND CONSTRUCTION OF A 50M SINGLE SPAN ULTRA-HIGH PERFORMANCE DUCTILE CONCRETE COMPOSITE ROAD BRIDGE

Yen Lei Voo[1], Patrick C. Augustin[2], Thomas A. J. Thamboe[3]

[1] Dura Technology Sdn Bhd, Malaysia.
[2] Perunding Faisal, Abraham dan Augustin Sdn Bhd, Malaysia.
[3] Endeavour Consult Sdn Bhd, Negeri Sembilan, Malaysia.

*Corresponding E-mail : *dura@dura.com.my*

ABSTRACT

A single span 50m long prestressed road bridge was constructed under Public Works Department in the State of Negeri Sembilan, Malaysia contract recently. The bridge was constructed at a small village, Kampung Linsum, crossing a river, Sungai Linggi. To date, this bridge is the Malaysia first and may also be the world longest composite road bridge which made from ultra-high performance ductile concrete (UHPdC). This paper presents the feature of the UHPdC precast girder; brief in-sight of the manufacturing of the girder; the construction sequence of the bridge; the design method and lastly the environmental impact calculation. The midspan deflections of the bridge at different construction history were compared against the collected field data and it showed that the calculated values generally agree well with the field data.

Keywords: *Ultra high performance, Bridge, Prestressed, Ductile.*

1.0 PROJECT BACKGROUND

The Public Works Department is the first to use ultra-high performance ductile concrete (UHPdC) in a bridge girder. The road bridge was completed in January 2011 (see Figure 1). The bridge was constructed using a single U-trough girder 1.75m deep, 2.5m wide at the top, topped with a 4m wide cast in-situ reinforced concrete deck 200mm thick. The UHPdC girder ends was encased in normal strength concrete abutments at the bridge site and made integral with the abutment seating. The girder was built without any conventional shear reinforcement as the UHPdC had considerable flexure and shear capacity. The UHPdC, with the trade name "DURA®" was supplied by Dura Technology Sdn. Bhd. It has achieved up to 180 MPa compressive strength and 30 MPa of flexural strength.

Figure 1: 50m single span UHPdC road bridge crossing Sungai Linggi, Negeri Sembilan.

2.0   DESIGN METHOD

2.1   BRIDGE LAYOUT

Figure 2 presents the general layout of the bridge. The transverse width of the bridge is 4m. The bridge is simply supported over a supporting length of 49.5m.

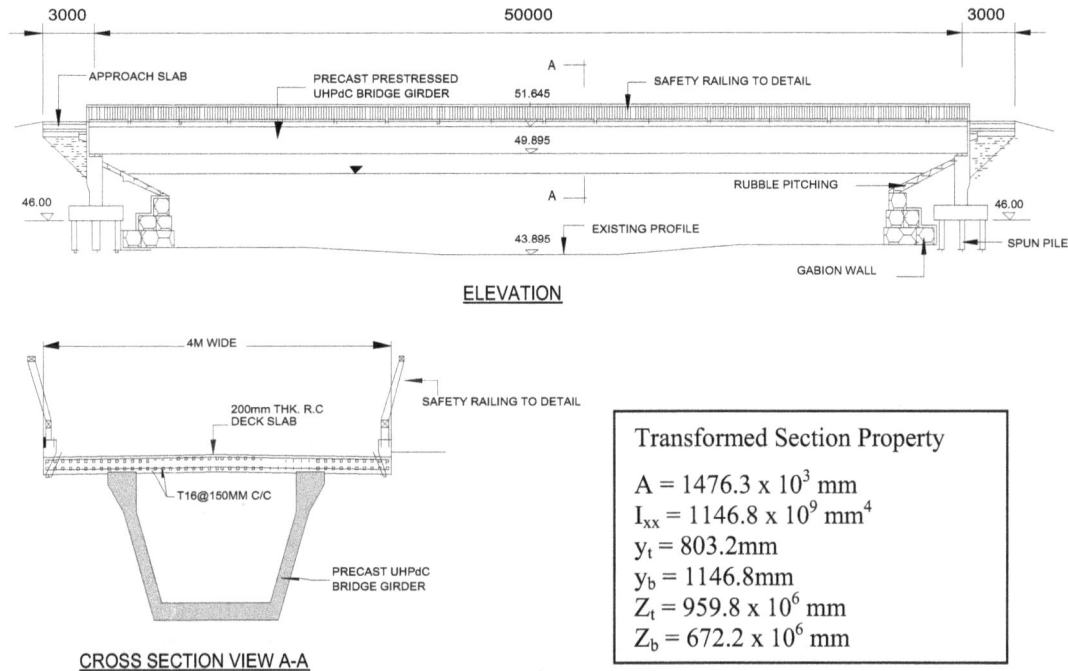

ELEVATION

CROSS SECTION VIEW A-A

Transformed Section Property

$A = 1476.3 \times 10^3$ mm
$I_{xx} = 1146.8 \times 10^9$ mm^4
$y_t = 803.2$mm
$y_b = 1146.8$mm
$Z_t = 959.8 \times 10^6$ mm
$Z_b = 672.2 \times 10^6$ mm

Figure 2: Layout of bridge.

2.2   SPECIFICATIONS OF BRIDGE

The specifications of the concrete bridge are as follows:
- Design life of structure: 120 years
- Number of nominal carriageway: 1
- Design traffic load: HA loading or 30 units HB loading (BS5400.2[7])
- Superstructure: Precast girder composited with 200mm thick in-situ Grade40 R.C. deck
- Bridge length: Single span of 50m
- Supported length = 49.5m
- Overall bridge width: B = 4m

2.3   LIMIT STATE DESIGN

2.3.1 GENERAL

- The bridge assumes to have relative humidity of 90% and average temperature of 30☐C (this information was used for time-effect analysis)
- The strands used are 7-wire stress-relieved type and has a diameter of 15.2mm, which come with a guarantee breaking load of 260kN per strand and modulus of elasticity of 195GPa. All the tendons were stressed to 75% of its break load. The immediate losses during stressing are taken as 5%. The relaxation of tendons at different time t can be

calculated according to clause 3.3.4 of AS3600[8]. The tendons stress limit at SLS shall not exceed 70% of its characteristic tensile strength (i.e. 1302MPa).

- The type of reinforcement used has yield strength and breaking strength of 410MPa and 460MPa, respectively. The modulus of elasticity is taken as 200GPa. The reinforcement stress limit at SLS shall not exceed 80% of its yield strength (i.e. 328MPa).
- Grade40 concrete assumed to have characteristic compressive of f_{ck} = 40MPa and tensile strengths of $f_{tk} = 0.36\sqrt{f_{ck}}$ = 2.3MPa. The shrinkage and creep models are as per the requirement of AS3600[7]. The basic creep coefficient is taken as $\phi_{cc,b} = 2.8$.

- The allowable deflection limit due to live load (i.e HA loading in this case) is taken as L/600 according to AS5400.2[9] bridge code. Therefore, the maximum deflection shall not be greater than 82.5mm.
- HA live load is used for stress limit and deflection check at SLS.
- HB live load is used as the most adverse strength criteria at ULS.

2.3.2 UHPDC MATERIAL PROPERTIES

The characteristic compressive strength of the UHPdC is taken as f_{ck} = 150MPa. Additionally, it is possible to take into account the tensile strength of the concrete as UHPdC is superior in its fracture property. The characteristic tensile strength can be taken as f_{tk} = 10.69MPa. The modulus of elasticity is taken as E_0 = 46.5GPa.

Conventional shrinkage and creep models based on standards/codes are not available for UHPdC as it is a relatively new material. Therefore, the shrinkage and creep models used for UHPdC are based on experimental data. Knowing that the post-production shrinkage and creep are minimal, in the calculation follows, the total shrinkage of 1000με (with early autogenously shrinkage as high as 500με to 600με) is assumed to be all undertaken after the steam curing. Therefore the post-production shrinkage is considered to be negligible (i.e. $\varepsilon_{sh}(t)$ =0).

The creep model used herein is the regression fit from the experimental work conducted in University of New South Wales on Dura®-UHPdC which has similar curing method as described in Section 3.3 (see Figure 4). Equation 4 as suggested by Voo and Foster[10] is used to model the creep coefficient $\phi_{cc}(t)$ of UHPdC at any time t, where $\phi_{cc,b}$ is the basic creep coefficient of the UHPdC (which is the mean value of the ratio of final creep strain to elastic strain for a specimen loaded at 28 days under a constant stress of 0.4f_{cm}) and may be taken as and $\phi_{cc,b}$ =0.20.

2.3.3 DESIGN ACTIONS

The design loadings, bending moment and shear force values are presented in Table 1.

Table 1: Design bending moments and shear forces.

	SLS		ULS	
	Moment (kNm)	Shear Force (kN)	Moment (kNm)	Shear Force (kN)
SW of U-girder (G₁= 22.9kN/m)	7014	600	8066	690
SW of Deck (G₂= 20.17 kN/m)	6178	504	7105	580
SW of Railing (G₃= 0.5 kN/m)	153	12.5	176	14.4
Live Load 1 (HA)	10824	735	-	-
Live Load 2 (30 units HB)	-	-	16263	1350
Total	**24169**	**1852**	$\mathbf{M_{Ed} = 31610}$	$\mathbf{V_{Ed} = 2634}$
Notes: The partial factor for UHPdC girder, RC slab and railing taken as γ_{fL} = 1.0 for SLS and γ_{fL} = 1.5 for ULS. The partial factor for HA live load is taken as γ_{fL} = 1.2 for SLS and γ_{fL} = 1.5 for ULS. The partial factor for 30 units HB live load is taken as γ_{fL} = 1.1 for SLS and γ_{fL} = 1.3 for ULS.

2.3.4 SECTION PROPERTIES

The effective flange width calculated as the full width for both SLS and ULS analysis. The cross section detail of the U-girder is presented in Figure 3. The U-girder consists of two slender vertical webs, each designed as a thin membrane element of 150 mm thick. The transformed sectional properties of the girder/composite bridge for the time-effect analysis are used herein corresponding to different load history of the bridge.

2.3.5 SERVICEABILITY LIMIT STATE (SLS)

The authors use the well established *Age Adjusted Effective Modulus Method* (AAEMM) by Gilbert and Mickleborough[11] to model the time-effect behaviour of the UHPdC composite bridge for a period of 30 years. The authors believe this method gives the most accurate prediction of the overall behaviour of the composite bridge at different load history during construction and during in service. Results on stresses, strains and deflections at the midspan are presented in Table 2.

In general, the stress levels for the concretes, tendons and reinforcements were confirmed to be below the specified stress limits. Calculation shows under the sustained permanent loadings for a period of 30 years, the prestressing strands will undergo maximum time-effect losses of 11.5% and 18% for the bottom tendons and top tendons, respectively. Also it has been observed that the resultant stresses of the steel reinforcements at the deck increases with time, from -20MPa to -100MPa, which indicates the inevitable creep and shrinkage behaviour of the normal strength concrete transfers significant amount of stress to the steel reinforcement.

Of particular interest, the AAEMM predicted deflection values are compared against the collected field data. Comparison show the AAEMM method generally is able to capture the overall deflection behaviour of the composite bridge at different load history during construction. AAEMM predicts the composite bridge will have a final sag deflection of 56mm after the 2nd stage PT, and the bridge shall bounce back to another 25mm after 30 years.

Table 2: Stress/stains and deflection at the midspan of the bridge (at SLS).

	Event	1st Stg. PT		Add RC Deck		2nd Stg. PT		HA Loading
	Days	28	57	57	71	71	10950	Infinity
	Analysis Type	I	T	I	T	I	T	I
	Composited?	No	No	No	Yes	Yes	Yes	Yes
Stress (MPa)	Slab Top	-	-	-	0.1	0.7	-2.2	-9.5
	Slab Bottom	-	-	-	0.2	-0.4	-2.6	-8.4
	Girder Top	-15.6	-15.8	-34.7	-34.7	-35.8	-24.1	-32.6
	Girder Bottom	-10.5	-9.6	4.2	4.2	-13.5	-13.7	2.4
	Top Reo.	-	-	-	-15	-13	-76	-121
	Bottom Reo.	-	-	-	-14	-16	-79	-119
	Top Strand	1231	1181	1110	1097	1088	1011	981
	Bottom Strand	1250	1208	1258	1262	1192	1106	1168
Strain ($\square\square$)	Slab Top	-	-	-	-78	-59	-369	-612
	Slab Bottom	-	-	-	-	-	-	-
	Girder Top	-335	-401	-807	-	-	-	-
	Girder Bottom	-226	-245	53	86	-294	-658	-311
Curvature ($10^{-6} \times mm^{-1}$)		0.0620	0.0890	0.4916	0.5484	0.3443	0.3168	0.6187
Theoretical Midspan Deflection (mm)		**-4.4**	**-0.1**	**103**	**118**	**56**	**31**	**106**
Field Measured Midspan Deflection (mm)		**-10**	**0**	**103**	**130**	**70**	**N/A**	**N/A**

- I = Instantaneous analysis   T = Time-effect analysis
- Grade40 Concrete Stress Limits = -040 f_{ck} = -16 MPa (in compression) and 2.5 MPa (in tension)
- UHPdC Stress Limits = -0.60 f_{ck} = -90 MPa (in compression) and 5 MPa (in tension)
- Reinforcing Steel Stress Limit = 0.8 f_{yk} = 328 MPa
- Prestress Strand Stress Limit = 0.70 f_{pk} = 1302 MPa

The instantaneous deflection at midspan can be calculated as the superposition of the UDL part and the KEL point load of the HA loadings. Therefore the instantaneous deflection at the midspan due to HA loading is calculated to be 75mm, which is less than the allowable deflection limit of 82.5mm. Therefore the section has sufficient stiffness to pass the deflection criteria.

2.3.6 ULTIMATE LIMIT STATE (ULS)

The calculation of the design moment resistance (M_{Rd}) of UHPdC composite bridge is no difference from conventional concrete bridge where simple beam theory can be used. In this case, the critical design moment is assume to locate at a distance $\pm$ 4 m from girder midspan as that section is a joint (i.e. joint 3 of Figure 3). The joint is assumed to have no residual tensile stress during ultimate stage. The width of the R.C. deck is taken as the full width of 4m. By equating the compressive and tensile forces through the cross-section, the neutral axis depth (d_n) is found to be in the precast U-girder with $d_n = 233.7mm$. From internal forces equilibrium and taking moment about the top extreme fiber, the ultimate moment capacity can be calculated as M_u = 47279kNm. Using a member reduction factor of $\phi = 0.8$, the design moment resistance can be taken as $M_{Rd} = \phi M_u = 0.8 \times 47279 = 37823$ kNm $> M_{Ed} = 31610 kNm$, which is greater than the design moment effect. Therefore the section has sufficient strength in flexure.

Since no stirrup is provided at any part of the UHPdC girder, the design shear resistance ($V_{Rd} = V_{yd}$) shall be set to either the design shear capacity of the web region as specified in the of Guidelines for UFC[1] or the shear transfer capacity of a dry keyed-joint specified in the experimental finding given in Voo[12], whichever is smaller.

From clause 6.3.3 of Guidelines for UFC[1], the design shear resistance can be calculated using Equation 1.

$$V_{yd} = V_{rpcd} + V_{fd} + V_{ped} \qquad \text{(Eq.1)}$$

where V_{rpcd} is the design shear capacity of a linear member that has no shear reinforcement bar, except the capacity provided by fiber reinforcement and is determined by:

$$V_{rpcd} = 0.18\sqrt{f'_{cd}}\, b_w\, d/\gamma_b = 0.18\sqrt{115.4} \times 300 \times 1728/1.3 = 771kN \qquad \text{(Eq. 1.1)}$$

where $b_w = $ *the width of the web* $= 2x150 = 300mm$;

$d - $ *effective depth* $= [(1950 - 100)x57 + (1950 - 350)x54]/(57 + 54) = 1728mm$;

$f'_{cd} = $ *design compressive strength* $= f_{ck}/\gamma_c = 150/1.3 = 115.4MPa$

$\gamma_c = $ *material reduction factor* $= 1.3$ and

$\gamma_b = $ *member reduction factor* $= 1.3$.

The term V_{fd} is the design shear capacity provided by the fiber reinforcement, which is determined by the following equation:

$$V_{fd} = (f_{vd}/\tan\beta_u)b_w\, z/\gamma_b = [8.22/\tan(30°)] \times 300 \times 1503/1.3$$
$$= 4938\ kN \qquad \text{(Eq. 1.2)}$$

where $f_{vd} = f_{tk}/\gamma_c = 0.9 f_{spk}/\gamma_c = 0.9 \times (16 - 1.65 \times 2.5)/1.3 = 8.22 MPa$ is   the design average tensile strength perpendicular to diagonal cracks of UHPdC, f_{tk} is the characteristic tensile strength of UHPdC in uniaxial tension, f_{spk} is the characteristic split cylinder strength of the UHPdC (refer to Table 4). According to AS3600[8] the tensile strength can be approximated as $f_{tk} = 0.9 f_{spk}$.

The term β_u is the angle between the member axis and a diagonal crack and it shall not less than 30 degree.

$$\beta_u = \frac{1}{2}\tan^{-1}\left(\frac{2\tau}{\sigma'_{xu} - \sigma'_{yu}}\right) - \beta_o \geq 30° \qquad \text{(Eq.1.2.1)}$$

where σ'_{xu} and σ'_{yu} are the applied average compressive stress along and perpendicular to the member. In this case σ'_{xu} is the average effective prestressing stress of the U-girder after time-effect losses. From Table 2, the average longitudinal stress can be taken as $\sigma'_{xu} = (-24.1-13.7)/2 = -18.9$ MPa and the perpendicular stress is taken as $\sigma'_{yu} = 0$.

The symbol $\beta_o = 5°$ is the angle formed by a diagonal crack and a line at 45° from the member axis, where it is not subjected to axial force.

The term τ is the average shear stress calculated from design shear force therefore it can be determined as:

$$\tau = (V_{Ed}/b_w\,d) = (2634\times10^3)/(300\times1728) = 5.08\ MPa. \qquad \text{(Eq. 1.2.2)}$$

Therefore, $\beta_u = \frac{1}{2}\tan^{-1}\left(\frac{2\tau}{\sigma'_{xu} - \sigma'_{yu}}\right) - \beta_o = \frac{1}{2}\tan^{-1}\left(\frac{2\times5.08}{18.9-0}\right) - 5 = 9.13°$. Since the value of β_u

shall not less than 30 degree. So in this calculation, $\beta_u = 30°$. The term z is the distance from the location of compressive stress resultant to the centroid of tension steel, which may generally be set to $z = d/1.15 = 1728/1.15 = 1503\ mm$.

The term V_{ped} is the vertical force from the tendon component, which is taken as $V_{ped} = 0$. Finally the design shear resistance is calculated as:

$$V_{Rd} = V_{yd} = V_{rpcd} + V_{fd} + V_{ped} = 771 + 4938 + 0 = 5709\,kN > V_{Ed} = 2634\ kN \qquad .$$

Therefore the section has adequate shear resistance for the design shear force.

The design shear capacity corresponding to diagonal compression failure V_{wcd} may be calculated by:

$$\begin{aligned} V_{wcd} &= 0.84 f'_{cd}{}^{2/3} \sin(2\beta_u)\, b_w\, d/\gamma_b \\ &= 0.84\,(115.4)^{2/3} \sin(2\times30°)\times300\times1728/1.3 \qquad \text{(Eq. 2)} \\ &= 6876\ kN \gg V_{Ed} = 2634\ kN \end{aligned}$$

Therefore web shear crushing is not critical.

In the following, the shear strength of the dry keyed-joint will be calculated using the experimental and analytical finding from Voo[12]. The principle of Mohr Circle was used to predict the shear strength of the dry joint by using the minor principal strength σ_{11} as the failure criteria (noted in this case $\sigma_{11} = f_{vd}$ of Eq. 1.2). The design shear resistance at the joint ($V_{j,Rd\ B}$) is taken as the superposition of the frictional force from the smooth-matched surface and the shear force contribution from the shear keys. Thus the design shear resistance of the joint can be written as:

$$V_{j,Rd} = V_{smd} + V_{kd} \qquad \text{(Eq. 3)}$$

where V_{smd} is the frictional force results from the compressive normal stress (i.e. $\sigma_n = -18.9$ MPa) due to prestressing, thus it can be expressed as:

$$V_{smd} = \mu A_{sm}\,\sigma_n/\gamma_b = 0.366\times405,000\times18.9/10^3/1.3 = 2155\ kN \qquad \text{(Eq. 3.1)}$$

where $A_{sm} = [(1750 \times 300) - (6 \times 100 \times 200)] = 405{,}000 mm^2$ is the area of the smooth section of the joint; γ_b is the member reduction factor which is set to $\gamma_b = 1.3$ for ULS according to Guideline to UFC[1]; μ is the friction coefficient which has calibrated against experimental specimens and it can be expressed as:

$$\mu = -0.0105\,\sigma_n + 0.5646 = -0.0105\,(18.9) + 0.5646 = 0.366 \qquad \text{(Eq. 3.1.1)}$$

The term V_{kd} is taken as the area of the key base times its maximum sliding shear strength (τ_{xy}), thus it can be written as:

$$V_{kd} = A_k\,\tau_{xy}\,/\,\gamma_b = 120{,}000 \times 14.93\,/\,10^3\,/\,1.3 = 1378 kN \quad \text{for ULS} \qquad \text{(Eq. 3.2)}$$

where $A_k = (200 \times 100) \times 6nos. = 120{,}000 mm^2$ is the total area of the shear key base and, the sliding shear strength is expressed as:

$$\tau_{xy} = \sqrt{\left(\sigma_{11} + \frac{\sigma_n}{2}\right)^2 - \left(\frac{\sigma_n}{2}\right)^2} = \sqrt{\left(8.22 + \frac{18.9}{2}\right)^2 - \left(\frac{18.9}{2}\right)^2} = 14.93 MPa \qquad \text{(Eq. 3.2.1)}$$

where, $\sigma_{11} = f_{tk}\,/\,\gamma_c = 0.9 f_{spk}\,/\,\gamma_c = 0.9 \times (16 - 1.65 \times 2.5)\,/\,1.3 = 8.22 MPa$ is the design tensile strength of the UHPdC at ULS which can be taken as the split cylinder tensile strength and $\gamma_c = 1.3$ is the material reduction factor.

Finally, $V_{j,Rd} = 2155 + 1378 = 3533 kN > V_{Ed} = 2634\ kN$. The shear resistance at the joint level is smaller than the shear resistance of the monolithic section (i.e. $V_{j,Rd} < V_{Rd}$), therefore the shear resistance at the keyed-joints governed.

3.0   BRIDGE CONSTRUCTION SEQUENCE

This section illustrates the construction sequence of the bridge. In brief the construction sequence can be summarized as:

Step	Activity	Date	Days
1	Fabrication of the UHPdC U-girder segments	Mid Oct 2010	-
2	Transportation of segments to job site	15th Nov 2010	20
3	Assembling of segments	16th Nov 2010	21
4	First stage post-tensioning	23rd Nov 2010	28
5	Launching of U-girder to abutments	3rd Dec 2010	38
6	Casting of in-situ RC deck	20th, 22nd Dec 2010	55, 57
7	Second stage post-tensioning	5th Jan 2011	71
8	Casting of the composite bridge to the abutment	13 Jan 2011	79

3.1   UHPDC U-GIRDER DETAIL

The precast girder consists of a total of seven segments, which consists of five standard internal segments (IS) each 8m long that weighed 18 tons, and two end standard segments (ES) each 5m long that weighed 15 tons (see Figure 3). Unlike conventional precast concrete girders, the UHPdC girder does not have vertical shear link in its thin webs. The only conventional reinforcements used are the bursting reinforcement at the anchorage zone, lifting reinforcement at the tendon deflector positions, and horizontal shear reinforcement at the top flanges where connection with the RC deck is required.

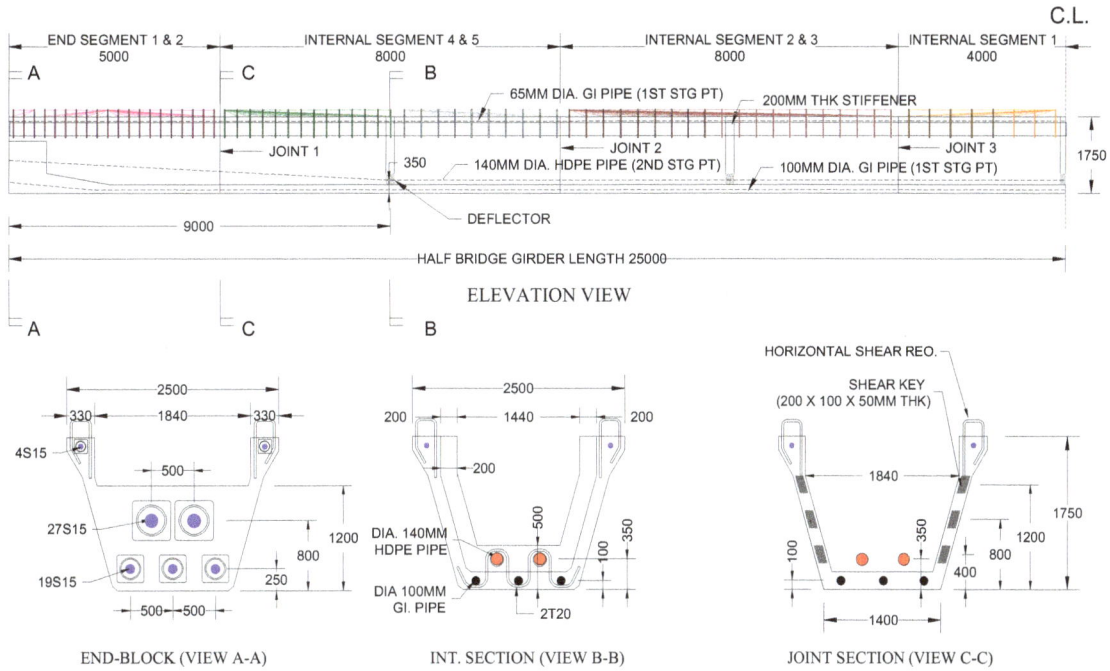

Figure 3: Detail of UHPdC girder.

3.2   MIX DESIGN OF UHPDC

The components of UHPdC are ordinary Portland cement, micro-silica, fine sand (with granular size less than 1mm), water, steel fibers and a high-range water reducing agent. In order to achieve the required performance of UHPdC, powder materials and fine aggregates are blended or proportioned to an adequate particle size distribution in order to maximize the density or compactness. Table 3 presents the general mix design used in the U-girder. Two types of steel fibers were used, that is (i) straight steel fiber with diameter and fiber length of 0.2mm and 20mm respectively; and (ii) end-hooked steel fiber with diameter and fiber length of 0.3mm and 25mm respectively. All this steel fiber has tensile strength beyond 2300MPa. The high-range water reducing agent used is a poly-carboxylate-ether based superplasticizer and no recycled wash water was used in the mixing.

Table 3: Mix design of standard DURA®-UHPdC (quantity in kg/m³).

Ingredient	Mass (kg/m³)
UHPdC Premix (Cement, micro-silica and fine sand)	2100
Superplasticizer	36
High strength steel fibers (>2300 MPa)	157 (2% by vol.)
Free water	144
3% moisture	30
Targeted W/B ratio	0.15
Total air voids	< 4%

3.3   MECHANICAL PROPERTIES OF UHPDC

UHPdC is a new generation of ultra-high performance construction material suitable for use in the production of precast elements for civil engineering, structural and architectural applications. UHPdC is a highly homogenous cementitious based composite without coarse aggregates which can achieve compressive strengths of 160 MPa and beyond. Its unique blend of very high strength micro-steel fibers and cementitious binders with extremely low water content give UHPdC the extraordinary characteristics of unique mechanical strengths, ductility comparable to steel and durability comparable to natural rock. UHPdC is highly impermeable, that is, the coefficient of water permeability and the diffusion coefficient of chloride ion are about $1/10^6$ which is 1/300 of ordinary high strength concrete. The Guidelines for UFC[1], suggested that this composite material can be used for more than 100 years without special repairs or reinforcement.

Table 4 summarized the mechanical properties of the UHPdC used in the U-girder. Each segment was cast from different batch of concrete, therefore it is important to take control samples of all the segments. Table 4 shows that the UHPdC can achieve cube compressive strength (f_{cu}) between 80MPa to 100MPa after 1 day; and 170MPa to 190MPa after 28 days. The cube compressive strength was measured according to BS 6319-2[2] using at least six cube specimens with dimension of 100mm.

Three pieces of 100mm diameter by 200mm height cylinders were tested for modulus of elasticity (E_0) and the experimental result shows the UHPdC has an average E_0 value of 46.5GPa. The E_0 values were determined according to BS 1881:121[3] under force control rate of 20MPa/min. The longitudinal strains were capture using electrical strain gauges.

Four pieces of 100mm diameter by 200mm height cylinders were tested for split cylinder indirect tensile strength test according to BS:EN 12390-6[4]. The first cracking strength (f_{ct}) is defined as the stress level in the UHPdC associated with a point in the stress-strain (or displacement) curve where the assumption of linear elasticity is no longer applicable. The split cylinder tensile strength (f_{sp}) is the point where maximum stress developed in the UHPdC after the first crack has formed.

Flexural toughness test as per ASTM-C1018[5] was used to determine the flexural properties of the UHPdC. Un-notched specimen with 100mm square cross-section and span over 300mm were used. A pre-load of 10kN was applied to the specimens and then unload to zero. This process was used for five cycles and then the specimens were loaded with midspan displacement control rate of 0.25mm/min till the end of the test. The load will decrease gradually after the peak load had achieved (i.e. f_{cf}). Experimental result shows all the UHPdC control specimens exhibit displacement-hardening behaviour after the first cracking (f_{cr}) occurred at an approximate midspan displacement of 0.03mm-0.05mm. This displacement hardening behavior is the results of the steel fibers bridging the micro-cracks and limiting the cracks from propagation. Experimental result shows the high volume of steel fibers in the concrete mix helps to increase the fracture mechanics of the composite, thus improved overall flexural toughness indexes (i.e. I_5, I_{10} and I_{20}).

The creep behavior of UHPdC was conducted at the University of New South Wales, Australia, over a period of 365 days, as per the specification of AS1012.16[6]. Four pieces of 100mm diameter by 200mm high cylinders and were pre-loaded with a compressive stress of 64MPa ($0.4f_{cm}$). The tests were conducted in an environmentally controlled room at an ambient temperature of 25°C and relative humidity of 50%. Two unloaded specimens were also measured to determine the shrinkage component of the measured strains. The experimental creep strain data are converted into creep coefficients, defined as the ratio of the creep strain to the instantaneous elastic strain, and the results are presented in Figure 4. From the results, the following equation was obtained for the creep coefficient:

$$\phi_{cc}(t) = \frac{15t^{0.4}}{60 + t^{0.4}} \phi_{cc,28} \qquad (Eq.4)$$

where $\phi_{cc}(t)$ is the creep coefficient at time t, and $\phi_{cc,28}$ is the creep coefficient at 28 days. For these tests, $\phi_{cc,28} = 0.2$.

3.4   FABRICATION

Manufacturing of the U-girder began in mid Oct 2010. Four segments are formed cast whereas three segments were matched cast against the formed cast segments (refer to Figure 5).

All the segments were steam-cured for a period of 48 hours at 90 degree Celsius as recommended by Guidelines for UFC[1]. Manufacturing of the last segment (i.e. IS1) was completed at early November. The total weight of the full girder was recorded to be 120 ton.

Table 4: Mechanical properties of UHPdC.

Ref.	Segment	End	Centre						End
		ES1	IS4	IS2	IS1	IS3	IS5	ES2
	Cast Type	FC	MC	FC	MC	FC	MC	FC
[2]	$f_{cu, 1 day}$ (MPa)	85	82	80	89	99	82	100
	$f_{cu, 28 day}$ (MPa)	189	180	171	183	188	183	193
[3]	$E_{o, 28 day}$ (GPa)	-	-	-	-	-	-	46.5
[4]	$f_{ct, 28 day}$ (MPa)	-	-	-	-	-	6.6	-
	$f_{sp, 28 day}$ (MPa)	-	-	-	-	-	16.0	-
[5]	$f_{cr, 28 day}$ (MPa)	14.5	15	14.0	14.5	15.5	15.0	15.5
	$f_{cf, 28 day}$ (MPa)	31	30	32	30	32	29	33
	I_5	5.95	5.86	5.52	6.00	5.81	5.94	6.22
	I_{10}	13.4	13.9	12.5	13.4	13.7	13.7	14.4
	I_{20}	29.9	31.4	27.7	30.1	31.1	31.3	32.6

Notes: FC = formed cast;   MC = matched cast;   ES = end segment;   IS = internal segment

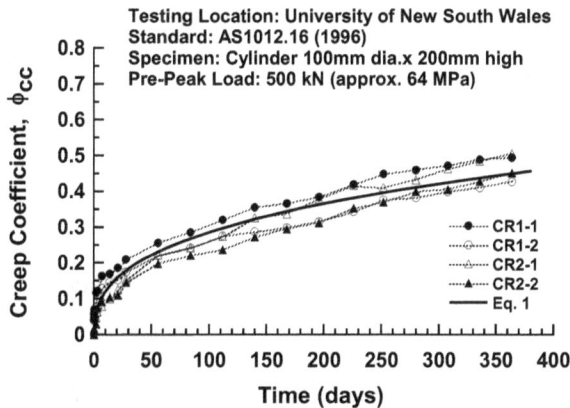

$$\phi_{cc}(t) = \frac{15t^{0.4}}{60+t^{0.4}} \phi_{cc,b} \quad ......[Eq.4]$$

Figure 4: Proposed creep Eq. 4 compare with experimental creep results on UHPdC.

3.5   ASSEMBLING OF U-GIRDER

A total of six 12m trucks were used to transport the seven segments to the job site. The segments were loaded onto the trucks on 15th November 2010 and on 16th November 2010 morning the segments arrived to the site and ready for assembling. Figure 6 shows a fully assembled U-girder at the job site. Due to the lightness of the segments, only one 45 ton mobile crane was used to align the seven segments.

3.6   FIRST STAGE POST-TENSIONING (PT)

First stage post-tensioning (PT) was carried out by Freyssinet PSC Malaysia on 23rd November 2010. Figure 7 shows the technician fitting the anchorage blocks for the three ducts@19S15 tendons (bottom row) and the two ducts@4S15 (top row). The central ducts (i.e. two ducts@27S15) were–for second stage PT. The bottom tendons were then stressed by a 7000kN capacity hydraulic jack. Both ends of the girder were stressed. At the end of the PT work, the midspan instantaneous hog deflection was measured to be approximately 10mm.

3.7   GIRDER LAUNCHING

Two units 160 tons mobile cranes were used to lift the UHPdC U-girder. In less than an hour, the U-girder was parked on one end of the steel framed transfer girder. Figure 8a shows the end of the U-girder was securely fastened on a trolley then gradually towed over the river. The girder was then securely positioned at the abutment of the bridge. The whole launching process took approximately 5 hours. At the end of the day, all the participants and witnesses were satisfied.

3.8   IN-SITU DECKING

Prior-to concreting, the level of the U-girder was measured and result shows the midspan deflection is close to 0mm (that is almost level). Figure 9a shows on 20th Dec 2010 (assumed day 1), the contractor concreted the first half portion of the deck. After the concrete was laid, the U-girder level was measured which showed a deflection of 25mm. After two days, the partially completed bridge had undergone further sag of another 25mm at the midspan due to the shrinkage effect from the RC deck. At this stage the net deflection is approximately 50mm. On 22nd Dec 2010 (day 3), the remaining half of the deck was concreted and the instantaneous midspan deflection recorded was 43mm, which give a total deflection of 93mm. On 24th Dec 2010 (day 5), further midspan sag deflection of 10mm was recorded which gave a total deflection of 103mm. Prior-to the second stage post-tensioning, on 4th January 2011 (day 16) the total midspan sag deflection of 130mm was recorded (refer to Figure 9b). These recorded deflection values are later compared against theoretical prediction as given in Table 2.

3.9   SECOND STAGE POST-TENSIONING

Second stage post-tensioning (PT) was carried out on 5th January 2011. The tendons were stressed by a 10000kN capacity hydraulic jack. Each duct was prestressed to a jacking force of 5265kN which gives a total prestressing force of 10530kN (refer to Figure 10a). At the end of the PT work, the internal ducts and deflectors were examined for defects. Examination shows the deflectors to be crack free (refer to Figure 10b). At this stage, the midspan instantaneous upward deflection was measured to be approximately 60mm (which result a net midspan deflection of 70mm).

3.10   LOAD TEST

Before the bridge was opened for the public use, the appointed consulting engineer from Public Works Department requested a load proof test on the bridge. Figure 11a shows an excavator weighing 22 ton was placed at the midspan of the bridge. The load test criterion is that under such static load, the bridge shall not deflect more than 16mm at the midspan and after the load is removed, the bridge shall have minimum 90% of recovery. In this test, a midspan displacement of 7mm was measured and after the excavator (i.e load) was removed the bridge has gained 100% recovery. Thus the authority has accepted the bridge deemed to comply. After a week, the bridge was open for public (refer to Figure 11b).

Figure 5: (a) Matched casting of internal segment (b) close up view.

(a)   (b)

Figure 6: One unit 45 tonnes mobile crane used to assembled the U-segment and (b) full U-girder assembled on site.

Figure 7: Stressing technician preparing for first stage post-tensioning.

(a)   (b)

Figure 8: (a) Launching of U-girder using steel A-frame transfer girder, (b) U-girder securely seated on abutments.

| (a) | (b) |

Figure 9: (a) Concreting of first half of the deck, (b) RC deck completed.

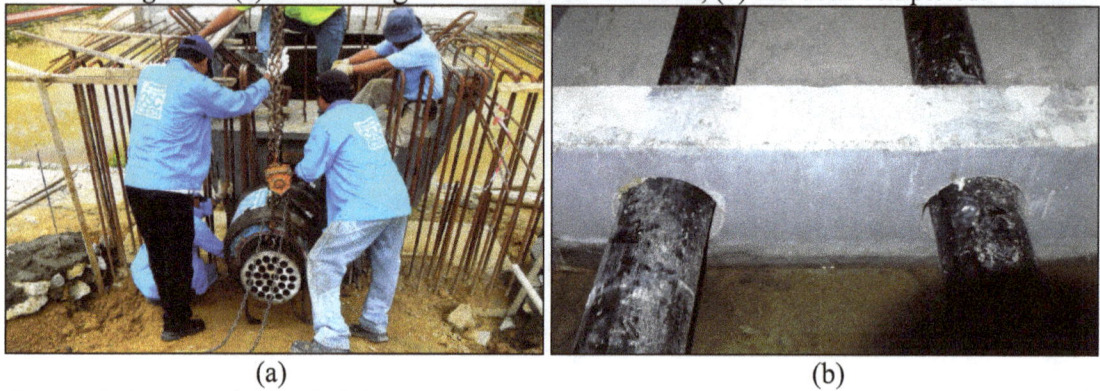

| (a) | (b) |

Figure 10: (a) Stressing technician preparing for second stage post-tensioning and (b) no sign of major cracking at the deviator.

| (a) | (b) |

Figure 11: (a) One 22 tonnes excavator place at the bridge midspan and (b) completed bridge open for traffic.

4.0   ENVIRONMENTAL IMPACT CALCULATION (EIC)

Infinitely, the design engineers of this bridge had proposed to use two structural steel welded girders for this bridge (see Figure 12). When UHPdC girder was proposed as an alternative, the consultants were amenable as the benefits, such as negligible maintenance, eco-friendly option, aesthetically pleasing and lower cost.

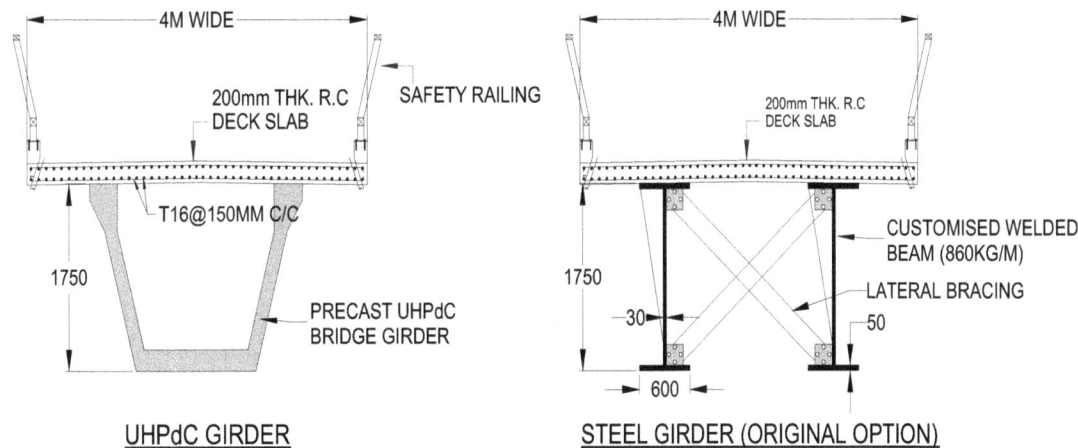

Figure 12: Comparison of UHPdC girder against steel girder composite bridge.

This section presents the environmental impact calculation (EIC) of the UHPdC composite bridge against the original steel beam composite bridge. Table 5 summaries the environmental data used in this comparative study, where detail on the derivation of the environmental impact data on the building material used can be obtained from Voo and Foster[10]. The table has been prepared to help to calculate the equivalent embodied energy (EE), CO_2 emission content and global warming potential (GWP) of particular concrete mix designs and materials.

In brief, Global Warming Potential (GWP) is a measure of how much a given mass of greenhouse gas is estimated to contribute to global warming over a given time interval. It is a relative scale which compares the gas in question to that of the same mass of CO_2. A 100-year of time horizon is most commonly used and it can be expressed as:

$$100\text{-year GWP} = CO_2 + 298\ NO_x + 25\ CH_4 \quad \text{(unit in ton of } CO_2 \text{ eq.)} \qquad \text{(Eq. 5)}$$

Table 6 summaries the material quantities and EIC of the two bridge designs. In the calculation of the material quantity, only the superstructure is considered herein. The amount of EE, CO_2 emissions and GWP are obtained from multiplying the amount of materials by the environmental data given in Table 5. Comparison of the EIC results is presented in Figure 13. In terms of material consumption, the UHPdC solution general consumed 14% more material (in term of weight) than the steel beam solution. In terms of environmental impact, the UHPdC solution is less environmental damaging with 66% less embodied energy and 57% less CO_2 emission. In terms of the 100-year GWP, the UHPdC solution gives a reduction of 52%.

Table 5: Environmental data.

	Units	Standard UHPdC (wt. 2% Steel Fiber)	Grade-40 (wt. 15% PFA)	Steel, Strand, Reo.
Density	kg/m^3	2400	2350	7840
EE	GJ/m^3	7.71	1.728	185.8
CO_2	kg/m^3	1065	297.5	17123
NO_x	kg/m^3	4.86	1.66	55.38
CH_4	kg/m^3	0.76	0.12	30.65
GWP	$kg\ CO_2\ eq.\ /m^3$	2532	795	34392
EE	MJ/kg	3.231	0.744	23.70
CO_2	kg/kg	0.446	0.128	2.184
NO_x	g/kg	2.035	0.714	7.064
CH_4	g/kg	0.318	0.052	3.909
GWP	$kg\ CO_2\ eq.\ /kg$	1.060	0.342	4.387

Table 6: Material quantities and environmental impact calculation (EIC).

		UHPdC (m^3)	Grade 40 Concrete (m^3)	Strands (ton)	Reo. (ton)	Steel (ton)	
No.	**UHPdC Composite Bridge**						
1	Precast U-girders	47.7	-	6.66	2.34	-	
2	RC deck	-	43.38	-	8.64	-	
	Sub-Total	47.7	43.38	6.66	10.98	-	**Total**
A	**Mass of material used (ton)**	114.48	101.9	6.66	10.98	-	**234.1**
B	**Embodied energy (GJ)**	368.0	74.97	157.8	260.23	-	**861.0**
C	**CO$_2$ (ton)**	50.80	12.91	14.55	24.0	-	**102.2**
D	**GWP (ton CO$_2$ eq.)**	120.8	34.49	29.22	48.17	-	**232.7**
No.	**Steel Welded Beam Composite Bridge**						
1	Steel welded beam	-	-	-	-	86	
2	Bracing (10% of beam)	-	-	-	-	8.6	
3	RC deck	-	43.38	-	8.64	-	
	Sub-Total		43.38	-	8.64	94.6	**Total**
A	**Mass of material used (ton)**	-	101.94	-	8.64	94.6	**205.2**
B	**Embodied energy (GJ)**	-	74.97	-	204.77	2242	**2521.8**
C	**CO$_2$ (ton)**	-	12.91	-	18.87	206.6	**238.4**
D	**GWP (ton CO$_2$ eq.)**	-	34.49	-	37.90	415	**487.4**

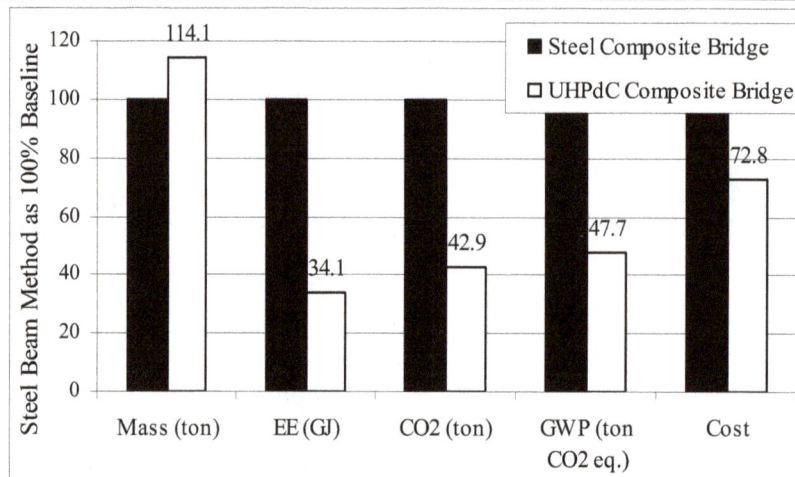

Figure 13: EIC comparison of UHPdC and steel composite bridges.

5.0   CONCLUSIONS

In January 2011, the Public Works Department had offered a tender for the construction of a single span 50m long bridge using an ultra-high performance ductile concrete composite design. It was optimized for a combination of structural, durability, sustainability and constructability aspects. To date, this bridge is Malaysia's first and perhaps the world longest road bridge with the main girders made from UHPdC. This paper presents the overview on the design and construction of the bridge. Age-adjusted elastic modulus method was used to predict the overall deflection value of the bridge corresponding to different load history during construction. Comparison shows the theoretical value generally is in well agreement with the field collected data. The bridge was then compared against conventional bridge design in term of environmental impact. In summary, the UHPdC design is confirmed to be a greener construction as the embodied energy content and CO$_2$ emission are approximately 66% and 57%, respectively less than of the conventional approach. In conclusion, UHPdC technology open the door for new design approach and it can make concrete structure more cost feasible, sustainable and environmental friendly.

REFERENCES

[1] *Recommendations for design and construction of ultra high strength fiber reinforced concrete structures (Draft)*, Concrete Committee of Japan Society of Civil Engineers, JSCE Guideline for Concrete, No. 9, ISBN: 4-8106-0557-4, September, 2006, 106 p.

[2] BS6319-Part 2: "Testing of resin and polymer/cement compositions for use in construction – method for measurement of compressive Strength", *British Standard*, British Standards Institution, 1983, 4 p.

[3] BS1881-Part 121: "Testing concrete – method for determination of static modulus of elasticity in compression", *British Standard*, British Standards Institution, 1983, 7 pp.

[4] BS:EN 12390-6: "Testing hardened concrete – tensile splitting strength of test specimens", *British Standard*, British Standards Institution, ISBN: 0 580 36606 5, 2000, 14 pp.

[5] ASTM-C1018: "Standard test method for flexural toughness and first crack strength of fiber reinforced concrete (using beam with third point loading)", *ASTM Standards*, ASTM International, United States, 1997, 8 pp.

[6] AS1012.16: "Determination of creep of concrete cylinders in compression", *Australian Standard*, Standards Australia, 1996, 8 pp.

[7] BS5400-Part 2: "Steel, concrete and composite bridges – part 2: Specification for loads", *British Standard*, British Standards Institution, ISBN 058009939 3, 1978, 71 pp.

[8] AS3600: "Concrete structures", *Australian Standard*, Standards Australia, ISBN: 0733793479, 2009, 198 p.

[9] AS5100-Part 2: "Bridge design – Part 2: Design loads", *Australian Standard*, Standards Australia, 2004, 71p.

[10] Voo, Y. L., and Foster S. J.: 'Characteristics of ultra-high performance 'ductile' concrete and its impact on sustainable construction', *The IES Journal Part A: Civil & Structural Engineering*, 3/3, 2010, p 168 – 187.

[11] Gilbert, R. I., and Mickleborough, N. C.: Design of prestressed concrete, 1st ed., Unwin Hyman Ltd.,1990, 504 p.

[12] Voo, Y. L.: Shear strength of steel fiber reinforced ultra-high performance ductile concrete dry joint for segmental girder, Technical Report No. TR-0006, Dura Technology Sdn Bhd, Malaysia, ISBN: 978-983-43785-5-4, June, 2010, 37 p.

LIST OF SYMBOL

A_k	total area of the base of shear keys (mm^2)
A_{sm}	total area of smooth section of the joint (mm^2)
b_w	the width of the webs (mm)
d	effective depth (mm)
d_n	neutral axis (mm)
$E_o,$	mean modulus of elasticity (GPa)
f_{cd}	design compressive strength (MPa)
f_{cf}	mean flexural strength / mean modulus of rupture (MPa)
f_{ck}	characteristic compressive strength (MPa)
f_{cm}	mean cylinder compressive strength (MPa)
f_{cr}	mean first bending cracking strength (MPa)
f_{ct}	mean first cracking strength (MPa)
f_{cu}	mean cube compressive strength (MPa)
f_{sp}	mean split cylinder tensile strength (MPa)
f_{spk}	characteristic split cylinder tensile strength (MPa)
f_{tk}	characteristic tensile strength (MPa)
f_{vd}	design average tensile strength perpendicular to diagonal crack (MPa)
I_5, I_{10}, I_{20}	mean flexural toughness indexes
L	nominal length of bridge (m)
M_{Ed}	design moment effect (kNm)
M_{Rd}	design moment resistance (kNm)
M_u	ultimate moment capacity (kNm)
V_{Ed}	design shear force effect (kN)
V_{fd}	design shear resistance provided by the fiber reinforcement (kN)
$V_{j,Rd}$	design shear resistance at joint region (kN)
V_{kd}	design shear resistance from the shear key of the joint (kN)
V_{ped}	design vertical force from the tendon component (kN)
V_{rpcd}	design shear force of linear member that has no shear reinforcement (kN)
V_{Rd}	design shear resistance (kN)
V_{smd}	design shear resistance from the frictional force of the joint results from the prestress normal stress (kN)
V_{wcd}	design shear capacity corresponding to diagonal compressive failure (kN)
V_{yd}	design shear resistance (kN)
t	time (days)
β_u	angle between the member axis and a diagonal crack (degree)
β_o	angle formed by a diagonal crack and a line 45° from the member axis, where it is not subjected axial force (degree)
ϕ	member reduction factor
$\phi_{cc}(t)$	creep coefficient at time t
$\phi_{cc,b}$	basic creep coefficient
$\phi_{cc,28}$	basic creep coefficient at 28 days
γ_b	member reduction factor
γ_c	material reduction factor
μ	friction coeffcieint
σ_{11}	minor principal strength (MPa)
σ_n	applied normal stress results from tendon component (MPa)
$\sigma,_{xu}$	applied average compressive stress along the member (MPa)
$\sigma,_{yu}$	applied average compressive stress perpendicular the member (MPa)
τ	average shear stress (MPa)
τ_{xy}	maximum sliding shear strength (MPa)

5

POTENTIAL USE OF MALAYSIAN THERMAL POWER PLANTS COAL BOTTOM ASH IN CONSTRUCTION

Abdulhameed Umar Abubakar[1], Khairul Salleh Baharudin[2]

[1]Research Student, Centre for Postgraduate Studies, Kuala Lumpur Infrastructure University College
[2]Assoc. Prof. Dept. of Civil Engineering & Infrastructure Technology, KLIUC, Selangor, Malaysia

*Corresponding E-mail : abdulhameedabubakar@rocketmail.com

ABSTRACT

As Malaysia focuses its attention to the call for a "greener" culture, so did the engineers and those in the scientific community especially the construction industry who is a major contributor to the depletion of green house gases. The engineering and construction community has now taken up the challenge for the use of "green and recycled by-products" in construction. One of those by-products is the Coal Bottom Ash (CBA) from thermal power plants that faces an increasing production running into hundreds of thousand tonnes in Malaysia alone, and its method of disposal is relegated to landfills alone with no other commercial usage. The construction industry is now forced to rethink on the utilization of the industrial by-products as supplementary materials due to the continuous depletion of natural aggregates in construction. A significant amount of research has been conducted elsewhere on CBA to ascertain its pozzolanic activity, compressive strength in concrete and mortar, durability, water absorption characteristics and density, in order to ensure its usage as a construction material. In this paper, a critical review of the strength characteristics of concrete and mortar as influenced by CBA as partial replacement of fine aggregate is presented based on the available information in the published literatures. Diverse physical and chemical properties of CBA from different power plants in Malaysia are also presented. The influence of different types, amounts and sources of CBA on the strength and bulk density of concrete is discussed. The setting time, workability and consistency as well as the advantages and disadvantages of using CBA in construction materials are also highlighted. An effective utilization of CBA in construction materials will significantly reduce the accumulation of the by-products in landfills and thus reduce environmental pollution.

Keywords: Coal bottom ash; By-products; Concrete; Mortar; Construction

1.0   INTRODUCTION

Ever since the Gospel of "Green Culture & Technology" was preached, it spread like "wildfire" consuming everything in its part to the extent that it created a conflict of interest between developed nations and their developing counterparts. Today, a connection has been "presumed" between Climate Change and the CO_2 content of the atmosphere which make it imperative to make use of by-products and low carbon products in the construction industry. The use of recycled & by-products is not only a desire but a necessity. It is interesting to look at this quotation in [1].

"Climate change has created a concern that pervades industrial and research thinking in the industrialized countries, since it presumes a casual relation between human industrial activities and changing climate. In other parts of the world, typically India and China, development of their economy seem to rank as high as concern over climate changes   caused by human activities. Thus a conflict is created."

Malaysia in its effort to promote National Green Technology Policy, has established a Green Technology Council that is aimed at coordinating ministries, agencies, the private sector and key stakeholders to lead to the implementation of the Green Technology Roadmap headed by the Prime Minister himself. In his budget speech of 2010, Prime Minister Najib Tun Abdul Razak announced measures to promote the use of environmentally friendly technology and resources in construction, including the establishment of a RM 1.5bn ($440m) fund for soft loans to companies that use and supply green technology. Thus, Malaysia is set to become the largest green construction sector in the South-east Asian region [2].

The above mentioned initiatives are as a result of the rising demand for energy from the economy and the call for a clean technology. In the electricity industry, Gas remains the main fuel source for the generation industry, but Coal is gaining favour. Gas-fired power plants make up almost 64% of all installed generating capacity; however, coal is growing in the generation fuel mix, rising from 11% in 2002, 26.7% in 2006 & 31% in 2008. The commissioning of two coal fired-power plants in 2009 means that coal use will continue to gain popularity [2]. The energy supplier has argued that coal is the only viable fuel option in terms of cost and supply. It is projected that the installed capacity on the coal power plants in the year 2010 will be 7,200MW (about 40% of the total) requiring about 22.5 million tonnes of coal, that is for 8,200MW capacities [3].

2.0   PRODUCTION OF COAL BOTTOM ASH

When pulverized coal is burned in a dry bottom boiler, about 80 to 90% of the unburned material or ash is entrained in the flue gas and is captured and recovered as Fly ash. The remaining 10 - 20% of the ash is dry Bottom ash, a dark grey, granular, porous, predominantly sand size material that is collected in water-filled hoppers at the bottom of the furnace. In wet bottom boilers, bottom ash is kept in a molten state and collected when it flows into the ash hopper below. The water in the hopper immediately fractures the molten material into crystallized pellets. In this case, the bottom ash is referred to as Boiler Slag (also known as "black beauty") a hard, black, glassy material. The remaining combustion products exit along with the flue gases, the figure below shows the combustion of coal to generate bottom ash in a thermal power plant.

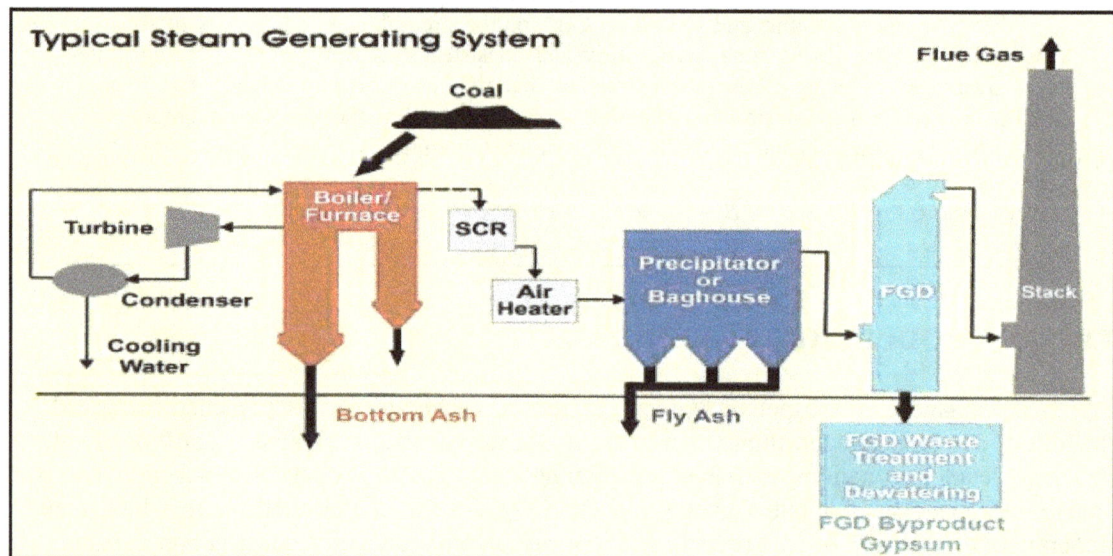

Figure 1: Typical thermal power plant & resulting waste generated

Source: NETL, 2006

The resulting coal ash generated is deposited either in a dry landfill over a vast area of land which is not possible in urban areas or deposited in an ash pond which also has its shortcomings. In Sejingkat Coal Fired Power Station, Kuching Sarawak, fly ash and bottom ash are deposited off into an 81,000m^2 area, 2.4m deep ash pond situated besides the power station. In fact, currently, there are two ash ponds with one of them fully utilized [4]. Tanjung Bin power station produce 180 tonnes per day of bottom ash and 1,620 tonnes per day of fly ash from 18,000 tonnes per day of coal burning alone [5]. The disposal of coal ash has reached an alarming proportion such that its use in construction is a necessity than a desire and if applied on a large scale, would revolutionize the construction industry, by economizing the construction cost and decreasing the ash content.

A report by American Coal Ash Association in 2008 presented in table 1 below showed that the utilization of coal bottom ash has been largely in the area of structural fill & embankment application. There is a relatively low application as blended cement due to its larger size particles & low pozzolanic properties until it has been ground to finer size particles [6] "... the compressive strength of mortar containing bottom ash at an early age depends on its fineness since the chemical composition of original & ground bottom ash are almost the same". Road base in pavement engineering, concrete, grout & aggregate has been the areas with the wider application.

Table 1: Utilization of bottom ash in tones

Application	Bottom ash (tonnes)
Concrete / concrete prod. /grout	720,948
Blended cement / raw feeder for clinker	610,194
Structural fill / embankment	2, 996, 388
Road bases / sub-bases	767, 013
Aggregate	727,048

Source: ACAA,2008

The methods of disposal have its disadvantages in the sense that when the pond or landfill site is not lined with concrete, heavy metals tends to leach into natural ground water and cause contamination. Recently, the disposal in an ash pond has come under heavy criticism with the 2008 collapse of one of the retaining walls at the Tennessee Valley Authority's Kingston power plant that released over a billion gallons of water & coal combustion ash slurry.

Results reported in literature are encouraging concerning the use of bottom ash as a partial or total fine aggregate replacement for natural sand [7]. It should also be noted that CBA is a highly porous material; therefore water absorption will have effect on the workability of the concrete. This paper intends to present discussion on the literature studies done on using CBA as a replacement for natural sand in concrete and mortar.

3.0   CHARACTERIZATION OF COAL BOTTOM ASH FROM DIFFERENT POWER PLANTS IN MALAYSIA

3.1   PHYSICAL PROPERTIES

Laboratory investigations carried out by [8] showed that bottom ashes have angular particles with a very porous surface texture. It sizes ranges from gravel to fine sand with very low percentages of silt-clay sized particles. Bottom ash is sand sized, usually 50-90% passing a 4.75mm (No. 4) sieve. It also has 10 - 60% passing a 0.42mm (No. 40) sieve, 0 - 10% passing a 0.075mm (No. 200) sieve, and a top size usually above 19mm. For categorization given in BS 882: 1992 based on percentage passing the 600µm sieve, between 55% to 100% would defined it as fine sand. The grading requirements for fine aggregates has been described into four zones in BS 882: 1973 and it was done based on the percentage passing the 600µm (No. 30 ASTM) sieve. The main reason for this is that large number of natural sands divides them at just that size, the grading above and below being approximately uniform [9]. Figure 2, shows a sample of coal bottom ash collected from a power plant.

Figure 2: Bottom ash (University of Kentucky, 2006)

From the study of [10] Wash Bottom Ash has percentage passing 600μm of 58.99%, therefore is considered as fine sand. The calculated fineness modulus of bottom ash is 3.65 which is more than 3.5 and is considered to be very coarse. It was reported by [11] when they performed sieve analysis on three different power plants bottom ash in West Virginia that the percentage passing the No. 4 (4.75mm) ASTM sieve size ranges from 90% - 52% which is in accordance with BS 882: 1992.

Table 2: Grain size analysis from different power plants

	Bottom Ash		
Sieve Size	Glasgow, WV	New Haven, WV	Moundsville, WV
38 mm (1-1/2 in)	100	99	100
19 mm (3/4 in)	100	95	100
9.5 mm (3/8 in)	100	87	73
4.75 mm (No. 4)	90	77	52
2.36 mm (No. 8)	80	57	32
1.18 mm (No. 16)	72	42	17
0.60 mm (No. 30)	65	29	10
0.30 mm (No. 50)	56	19	5
0.15 mm (No. 100)	35	15	2
0.075 mm (No. 200)	9	4	1

Source: Moulton & Lyle, 1973

Result of Sieve analysis by [8] showed that bottom ashes have angular particles with a very porous surface texture. It sizes ranges from gravel to fine sand with very low percentages of silt-clay sized particles. The ash is usually a well-graded material, although variation in particle size distribution may be encountered in ash samples taken from the same power plant at different times.

The figure (3) below is a plot of the grading analysis of three different power plants in Malaysia by [12]. Result from Tanjung Bin showed that it is distributed from fine gravel to fine sand, while Kapar and Manjung has a distribution from coarse-medium sand to fine sand, and majority of the sizes ranges between 10mm – 0.075mm. The work of [5] when they performed a sieve analysis on two bottom ash specimens, reported that the specimens were quite similar and exhibit well graded distribution. Their sizes ranging from fine gravel to fine sand sizes and the majority sizes occurred in a range between 20mm and 0.075mm.

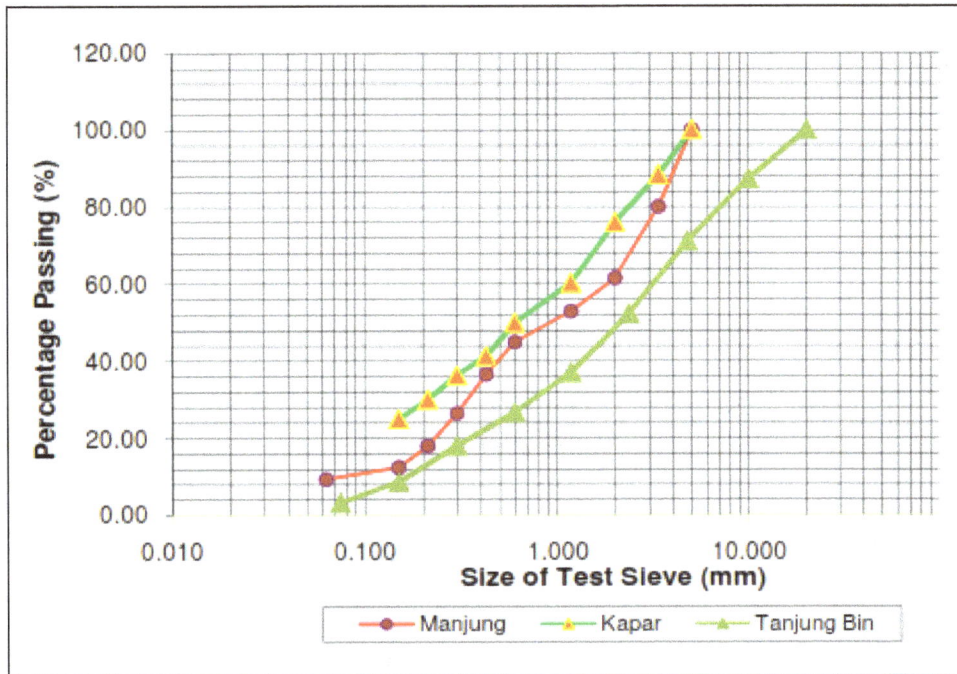

Figure 3: Sieve analysis of Malaysian power plants CBA
Source: Abdul Talib, 2010

Specific gravity of bottom ash is a function of chemical composition, with higher carbon content resulting in lower specific gravity. Coal Bottom ash with a low specific gravity, has a porous or vesicular texture, a characteristics of popcorn particles that readily degrade under loading or compaction [13]. Bottom ash with a porous or hollow ash may present a specific gravity as low as or even lower than 1.6 [14]. The range of the specific gravity of this bottom ash might be different depending on coal type, origin, size, handling, processing technique, boiler size, disposal and storage methods or other criteria [15]. All these factors mentioned have a commanding influence on the specific gravity property of bottom ash, as reported by [16], when they investigated the physical properties of bottom ash specimen taken from different disposal points in a disposal pond. "There is a difference of Gs between the sample taken from nearest to slurry disposal point and one taken farther away".

Kapar bottom ash seems to display lower specific gravity of 2.01 [12]. The lower specific gravity of the bottom ash is due to its chemical composition which is low in Iron oxide content and display of vesicular texture. The specific gravity value is directly proportional to iron content but for lime content greater than 15%, the value is more irrespective of the iron content [17]. Another factor that has an influence on specific gravity is the state of the material, the specific gravity of a dry bottom ash ranges from 2.0-2.6 with an average of 2.35 while wet bottom ash ranges from 2.6-2.9 with an average of 2.75 [14].

3.2 CHEMICAL PROPERTIES

The chemical analysis of CBA either using X-ray energy dispersive spectrometry (EDS) or X-ray fluorescence (XRF) will reveal that the main chemical compounds include Silicates (SiO_3), Aluminates (Al_2O_3) & Iron oxide (Fe_2O_3) with a host of other compounds in smaller percentages. The results of chemical composition analysis conducted on three thermal power stations bottom ash in Malaysia namely Kapar thermal power station in Selangor, Jimah Power plant in Negeri Sembilan and Tanjung Bin in Johor is presented in table 3 from independent researches.

Table 3: Chemical analysis results from different power plants in Malaysia.

	Bottom ash percentages		
	Muhardi et al., (2010)	Fawzan (2010)	Naganathan et al., (2012)
Chemical contents	Tanjung Bin Power Plant	Jimah Power Plant	Kapar power Plant
SiO_2	42.7	49.4	9.78
Al_2O_3	23.0	22.3	20.75
Fe_2O_3	17.0	13.7	37.1
CaO	9.80	9.0	11.1
K_2O	0.96	1.00	-
TiO_2	1.64	2.2	-
MgO	1.54	0.87	3.2
P_2O_5	1.04	0.65	-
Na_2O	0.29	0.13	-
SO_3	1.22	0.68	-
BaO	0.19	-	-
MnO	-	0.08	-
ZnO	-	-	1.8

The major components of the three thermal power plants in Malaysia studied by [5, 18, and 19] were Silica, Alumina & Iron oxide with percentage compositions of 9.78 - 49.4%, 20.75 - 23% & 17 - 37.1% respectively. The bottom ash used by [5,18] is a Class F because the sum of $SiO_2 + Al_2O_3 + Fe_2O_3$ exceeds 70% and according to ASTM C618 this can be attributed to the use of Bituminous or Anthracite Coal which produce low calcium content. The bottom ash studied by [19] is a Class C because the sum is less than 70% but greater than 50%. Class C is generated from the combustion of Lignite or Sub-bituminous coal with a high calcium content. Smaller percentages of potassium, magnesium & sodium are also present in Malaysian power plant bottom ash with traces of barium, manganese & zinc. The SO_3 content for the both Class F bottom ash of [5, 18] were 1.22% & 0.68% both of which are less than 2.5% specified by BS 3892: Part 1: 1993. The alkali K_2O & Na_2O which are insoluble residue were recorded as 0.96% & 1.00% and 0.29% & 0.13% respectively less than the 1.5% reported by ASTM C 618 – 94a. The result of Class C bottom ash of [19] did not present values for SO_3, K_2O & Na_2O respectively. A research conducted by [20] reported that "In bituminous coal, three major components 'SiO_2, Al_2O_3, Fe_2O_3' accounted for about 90% of the total components whereas lignite or sub bituminous coal ashes had relatively higher percentages of CaO, MgO & SO_3.

4.0   INFLUENCE OF COAL BOTTOM ASH ON THE MECHANICAL PROPERTIES OF CONCRETE & MORTAR

The strength of concrete is influenced by the volume of all voids in it and considering the fact that we are dealing with a very porous material (bottom ash), then it is only natural that we see a decrease in strength properties and unit weight of the resulting concrete. According to [6], "Bottom ash has a large particle size and a high porous surface, resulting in higher water requirement and lower compressive strength". The water retention capacity of bottom ash has also been highlighted by [21] & [22] that "additions increase the capacity of aerated block to retain water since Bottom ash is a porous material, thereby improving the moisture transport behavior within the block during fire". This implies that the water retention property of bottom ash is an advantage during fire functioning as a reservoir.

4.1   SETTING TIME & CONSISTENCY

Setting refers to the stiffening of the cement paste, a change from liquid to a rigid stage [9]. It is equally important to differentiate between setting and hardening which by definition refers to the gain of strength of a cement paste. A minimum time of 60 minutes is prescribed by ENV 197-1: 1992 for cements with strengths up to 42.5MPa and ASTM C 150-94 prescribes a minimum time for the initial set of 45 minutes using Vicat apparatus. The European Standards (EN 450-1, 2005) require the initial setting time of fly ash paste should be at most 120 minutes longer than that of the reference without ash.

Consistency on the other hand for any given cement, is the water content which will produce a paste of standard consistency [23].The test is also prescribed by EN 196-3: 1987. The water content of the standard paste is expressed as a percentage by mass of the dry cement, the usual range of values being in the range of 26-33%. According to [6], when they investigated the development of bottom ash as pozzolanic material by grinding the bottom ash to a smaller particle size, found that the normal consistency of the cement paste was 24.9% while that of the original and ground bottom ash was between 24.7%-25.3%. This is within the limit specified by EN 196-3: 1987.

Studies have identified that the addition of bottom ash to cement materials increases the initial and final setting time in relation to the reference mix [24]. This is due to the increase in quantity of water present in the mixes with bottom ash, resulting in the maintenance of a greater workability, consequently, increasing the time that the mix is in the fresh state. The use of original and ground bottom ash to replace Portland cement has being found to slightly retard the setting times of the paste as reported by [6]. They found that the setting times of cement paste were 112 & 180 minutes for the initial and final setting times respectively. The original and ground bottom ash initial setting times were between 122 minutes and 135 minutes while the final setting times were between 195 minutes and 210 minutes. They attributed the delay in the setting times to the replacement of bottom ash with cement which reduce the quantity of C_3S in the paste.

In [21], it was reported that the setting time of blocks made completely with and without recycled material. The initial and final setting times of the block without bottom ash and fly ash were 47 minutes & 291 minutes respectively while that of blocks with recycle material completely was 152 minutes and 1502 minutes for initial & final setting times respectively. The result showed that the initial setting time was longer by 105 minutes and therefore the limit of EN 450-1, 2005 is not exceeded.

In [25], it has been reported on the effects of natural Pozzolan on properties of cement mortars that "Setting properties of cement matrix were affected by natural Pozzolan ratio substituted for cement. Experimental results show a proportional delay in the initial set time, depending on the natural Pozzolan addition ratio". This implies that natural Pozzolan decrease rate of hydration by decreasing heat of hydration.

4.2   WORKABILITY & FLOW OF BOTTOM ASH

According to [9], defining workability as "ease of placement and the resistance to segregation" is too loose a description of this vital property of concrete. Workability should be defined as a physical property of concrete without reference to the circumstances of a particular type of construction. The most important factor affecting workability is the water content of the mix, if the water content and other factors are fixed, workability is governed by the maximum size of the aggregate, its grading, shape and texture.

Coal bottom ash has been reported to have a porous texture and angular shape, therefore it will require a considerable amount of water to produce a workable mix and bearing in mind that strength of concrete depends on the water-cement ratio, the higher the w/c ratio the lower the strength of the concrete. Its angular structure has been one of the factors that is restricting it application as reported by [26 - 27] "recycling of bottom ash in construction is restricted for the following reasons; the ash has angular structural characteristics, contains unburned carbon particles". In [28], they also concur that "A close examination of bottom ash reveals that it has a rough surface texture and is comprised of angular and porous particles. The irregular and angular shape and very porous surface necessitated the use of a higher water content to achieve the degree of lubrication needed for a workable mix"

With respect to mortar, the same observation was made that the flow of mortar decreased when bottom ash is used as fine aggregates. The flow of fresh mortar and workability of concrete are mainly related to size and shape of aggregate particles. The demand for the additional water is increased due to fine particle sizes, amounts and the internal water content of the bottom ash that is stored in the pores structures. Adequate considerations should be given to aspect of water demand in the mix design of bottom ash concrete

4.3   INFLUENCE OF STRENGTH AND BULK DENSITY ON CONCRETE & MORTAR

The mechanical strength of hardened cement is the property of the material that is perhaps most obviously required for structural use [9]. The strength of mortar or concrete depends on the cohesion of the cement paste, on its adhesion to the aggregate particles, and to a certain extent on the strength of the aggregate itself. In this section, the strength of concrete and mortar samples made with different percentages of coal bottom ash will be discussed.

According to [29], large size (greater than 6mm) bottom ash can be used as coarse aggregate and small size can be used as fine aggregate. Also, [30] showed that it is possible to manufacture lightweight concrete (LWC) with Saturated Surface Dry (SSD) in the range of 1560-1960kg/m^3 and a 28 day compressive strength in the range of 20-40 N/mm^2. The test was conducted in two series M (design strength of 20 N/mm^2) and H (design strength of 40 N/mm^2). For series H, compressive strength decreased at all ages, but for M, the decrease was only observed at 3 day strength. There was an increase in strength at 7 & 28 days when natural sand and coarse aggregates were replaced (see figure below).

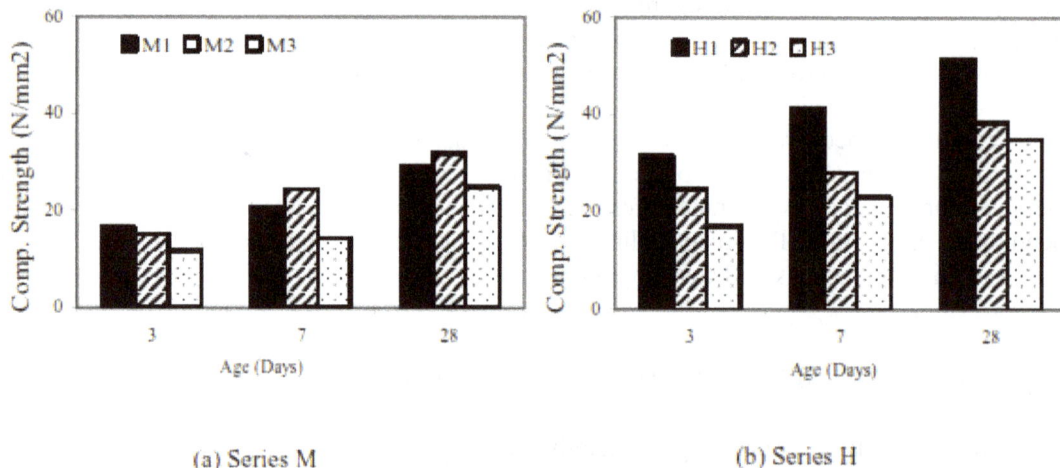

(a) Series M   (b) Series H

Figure 4: Compressive strength for M & H series

Source: Bai, Ibrahim & Basheer, 2010

M1, H1 – Control
M2, H2 – 100% OPC + 100% FBA + 100% LG
M3, H3 – 70% OPC + 100% FBA + 100% LG

They also observed that the density of hardened concrete at SSD condition measured at 28 days indicated a significant reduction in the density of the hardened concrete for both series. When they further replaced 30% of OPC with fly ash, the density of the resulting concrete was further decreased, this they attributed to the low density of fly ash compared to that of OPC.

Table 4: Density of hardened concrete at 28 days (SSD)

M1	M2	M3	H1	H2	H3
1977	1725	1559	2471	1952	1819

Source: Bai, Ibrahim & Basheer, 2010)

As Observed in [28], the strength development for bottom ash concretes follows the pattern of strength gain of the control concrete. However, as they noted, the gain was initially slower than that of the control sample; "….at the end of 60 days curing, an additional gain of 7% was recorded and after 90 day curing, a 13.7% increase in strength over that at 28day was observed". Also, [31] observed that the range of compressive strength of bottom ash mixture was between 4.2-12.5MPa compared to 15.9MPa of the control mix at the age of 3 days for CRT3 and 16.1-21.2MPa for CRT4. When the curing period was extended to 28 days CRT3 recorded 8.6-23.2MPa and control of 28.4MPa while CRT4 26.1-32.6MPa. At the age of 90 days, CRT3 was in the range of 12.5-25.7MPa with the control at 32MPa and CRT4 32.1-38.4MPa.

Table 5: Results of compressive strength & density at different ages of bottom ash

Concrete	Fresh density (kg/m^3)	Compressive strength (MPa)		
		3 days	28 days	90 days
0% CRT	2238	15.9	28.4	32.0
25% CRT3	2177	12.5	23.2	25.7
50% CRT3	2090	9.9	18.0	23.0
75% CRT3	1964	6.3	11.5	14.9
100% CRT3	1869	4.2	8.6	12.5
25% CRT4	2220	19.5	27.2	32.1
50% CRT4	2138	17.0	28.5	35.9
75% CRT4	2109	16.1	26.1	32.7
100% CRT4	2040	21.2	32.6	38.4

Source: Andrade et al., 2008

The work of [21] showed that the result of flexural strength for blocks with percentage replacements of bottom ash follows the same trend as that of the compressive strength. The bottom ash was replaced with fine aggregate in percentage replacements of 20 % & 30% respectively. The control block had a flexural strength of 4.4 MPa, while the bottom ash blocks had 3.3MPa & 3.1MPa respectively for 20 & 30 percent replacement levels, which is consistent with the result of the compressive strength where the control block had a strength that is twice that of the 20% bottom ash block. In general, the higher the bottom ash percentage in the block resulted in lower compressive and flexural strength developed as well as the density of the resulting blocks.

Table 6: Variation of density, compressive & flexural strengths of bottom ash blocks

Mixture	Density (kg/m^3)	Rc (MPa)	Rf (MPa)
H0 (control)	2090	22.7	4.4
H-BA20	1740	11.6	3.3
H-BA30	1660	8.2	3.1

Source: Arenas et al., 2011

In terms of the splitting tensile strength of concrete, [28] observed that no reduction in splitting tensile strength as compared to that of the equivalent conventional specimens as long as minimum cement content of $365kg/m^3$ is utilized. The inclusion of bottom ash had more influence on tensile strength than compressive resistance. This observation is in agreement with the original hypothesis of conventional concrete that states the use of alternate materials affects tensile strength differently than compressive strength.

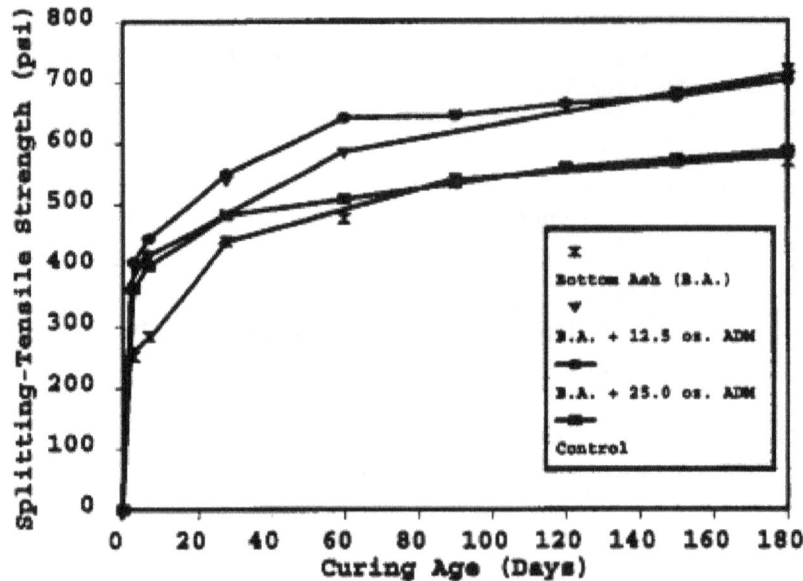

Figure 5: Splitting – tensile strength of concrete containing $600lb/yd^3$ $(365kg/m^3)$ cement

Source: Ghafoori & Bucholc, 1996

According to [28], they opined that "In as much as the bottom ash concrete displayed an identical, and in some cases superior, splitting-tensile strength compared to the control mixes, the use of a water-reducing admixture was not necessary for the improvement of the splitting-tensile strength of the bottom ash concrete." Their work consisted of the use of a water reducing admixtures in an attempt to reduce the water – cement ratio.

The strength of mortar composed of coal bottom ash has been found to be affected by a number of factors such as, curing period & the percentage composition. The amount of water has an influence on the strength properties of mortar and concrete. It was reported in [19] that "strength is mainly dependent on the mixing ratio and the water contained" when they investigated the development of brick using thermal power plant bottom ash & fly ash. If excessive water is used, then the strength will decrease [32]. They observed that the more cement used, resulted in high compressive strength, the strength was highest for a mix with water-cement (w/c) content of 412.0/208 and lowest for 386.7/79.4 which had the lowest cement used. The strength of bricks developed at 28 days ranged from 4.3MPa to 10.96MPa. According to [33], "the minimum strength for class 1 brick is 6.9MPa", therefore the bricks developed by [19] is comparable to that of normal clay bricks. They also noted that the density varied from 1500-1650kg/m^3, the relationship between fresh density and water-powder ratio (w/p) indicated that increasing the w/p ratio increases the density of the mix.

5.0   ADVANTAGES OF USING COAL BOTTOM ASH

The use of coal combustion by-products in construction has been shown to provide alternative solutions to the problems of global warming and the depletion of greenhouse gases as well as providing a sustainable future in the use of green and recycled products. The following are some of the benefits that would be derived from the sustained use of coal bottom ash in construction when proper and established standards have been set.

1. It is possible to produce lightweight concrete with a density in the range of 1560-1960 kg/m^3 and a 28 day compressive strength in the range of 20-40 N/mm^2 [30]. Though, the strength development is slow at the beginning but with extended curing days, maximum strength can be achieved.

2. Bottom ash may be used as a partial replacement of natural aggregates, with finer bottom ash used as sand. The percentage of bottom ash that can be used in a mixture composition depends upon its quality and required strength of the product [34].

3. Inclusion of bottom ash has a more pronounced influence on tensile resistance than on compressive strength, reduction of splitting tensile strength is hardly noticed, as long as a minimum cement content of 365 kg/m^3 is utilized [28].

4. Drying shrinkage decreased with an increase in bottom ash content. Concrete made from bottom ash exhibits a reduced drying shrinkage in comparison with that of the control samples [28].

5. Due to increased demand for mixing water, bottom ash mixture displayed a much higher degree of bleeding than the control concrete [28].

6. High fire resistance: for protection against fire, those materials that retain large quantity of water are more desirable, since when they are exposed to a fire, part of this water evaporates and is transported from the fire exposed surface to the interior of the material, where the water cools and condenses again [35].

7. Fly and bottom ashes increase the fire resistance of blocks, and are principally due to the wide evaporation plateau that those ashes incur as a result of increase water intake of the porous aggregates [21]

6.0   DISADVANTAGES OF USING COAL BOTTOM ASH

The use of coal bottom ash in concrete and mortar production has enormous advantages and potentials in the long run, but there are some shortcomings that have to be overcome.

1. The early strength development of coal bottom ash has been shown to be very slow at the beginning, but as the curing period is extended beyond 28days, a dramatic increase in strength is noticed [6].

2. Bottom ash mixtures display a lower modulus of elasticity than the control mixes. The empirical relationship between static modulus of elasticity, unit weight and compressive strength is slightly lower than that suggested by the American Concrete Institute (a=31.2) [28].

3. Due to high water absorption rate, angular shape and very porous surface of the bottom ash, higher water content is required to achieve the degree of lubrication needed for a workable mix. The increase water demand has a moderate effect on early-age characteristics of bottom ash concretes [28].

4. The inclusion of bottom ash has been shown to delay the setting time of the mixture with increase in percentage of bottom ash, the initial setting time is further delayed. This can be attributed to the reduction in the quantity of C_3S as a result of adding bottom ash and the amount of mixing water required to maintain a workable mix

7.0   CONCLUSIONS AND FUTURE RESEARCH ON COAL BOTTOM ASH

Ten years ago, statistics have shown that the coal consumption by thermal power plants in Malaysia was 11% of the total energy demand of Tenaga Nasional Berhad and it was estimated that by 2010, the energy demand and coal usage will jump to 40% out of which 15-20% of the by-product is bottom ash which at the moment has no commercial application but only left to litter landfills unlike its companion fly ash. The use of coal bottom ash in construction will serve to reduce carbon dioxide emission as a result of cement-based processes; imbibing the culture of green building and technology by a sector that is a major contributor of emissions and reduction of landfills and landfill cost to the government.

Available literatures have shown that the application of coal bottom ash by partially or fully replacing sand in concrete or mortar production in percentage replacements of about 10-30 percent have shown remarkable improvement with a small reduction of strength which can be overcome with a longer curing duration due to the delay of pozzolanic reaction of bottom ash until later ages. Also, contrary to the popular opinion that bottom ash is an inert material, it has been shown that with appropriate percentage replacements, bottom ash can perform better than the reference samples at later ages of curing period due pozzolanic reaction after the initial stages which is relatively slow.

Future research on the use of coal bottom ash in construction should focus on the following:

1. An established set of standards that spells out guidelines on its usage and regulates it if need be.
2. Long-term study on the effect of durability and strength properties of concrete and mortar using coal bottom ash is required.
3. The influence of constituent materials on the water absorption [19].
4. Chloride transport and Corrosion effect of coal bottom ash concrete.

REFERENCES

[1] Elkem, P. F., (2010) "Sustainable construction for the future: high performance concrete with microsilica." Concrete Technology Today.

[2] Malaysia 2010, The Report. Oxford Business Group. www.oxfordbusinessgroup.com

[3] Mahmud, H. O. (2003) "Coal- Fired Plant in Malaysia". The 15th JAPAC International Symposium. 19 September, 2003. Tokyo.

[4] Tsen, M. Z., (2008) "The properties of fly ash-based geopolymer mortar with potassium based alkaline reactor. Unpublished B.Eng Thesis, Curtain University of Technology.

[5] Muhardi, Marto, A., Kassim, K.A., Maktar, A.M., Lee, F.W. and Yap, S.L., (2010) "Engineering Characteristics of Tanjung Bin Coal Ash" Electronic Journal of Geotechnical Engineering. Vol. 15 pp. 1117-1129.

[6] Jaturapitakkul, C. and Cheerarot, R., (2003) "Development of bottom ash as pozzolanic Material". *J Mater Civ. Eng*, Vol.15 (1), pp. 48-53.

[7] Manz, O.E., (1997) "Worldwide production of coal ash and utilization in concrete and others Products" *Fuel*, Vol. **76** (8), pp. 691–696.

[8] Muhammad, M.S., (2010) "Strength characteristics of Compacted Residual Soil-Bottom Ash mixture" Unpublished B.Eng. Thesis. Universiti Teknologi Malaysia.

[9]Neville, A.M., (1995) "Properties of Concrete" Pearson education limited.

[10] Syahrul, M.H., Muftah, F. & Muda, Z., (2010) "The properties of special concrete using washed bottom ash (WBA) as partial replacement." International Journal of Sustainable Construction Engineering & Technology. Vol. 1, No. 2 pp. 65-76.

[11] Moulton & Lyle (1973) "Bottom Ash and Boiler Slag" International Ash utilization symposium. US Bureau of Mines, Information circular No. 8640, Washington DC.

[12] Abdul Talib, N.R., (2010) "Engineering characteristics of Bottom ash from power plants in Malaysia". Unpublished B.Eng. Thesis. UTM.

[13] Lovell, C.W., Huang, W.H. and Lovell, J.E., (1991) "Bottom ash as highway material" Presented at the 70th Annual Meeting of the Transportation Research Board, Washington, D.C.

[14] Kim, B.J., Yoon, S.M. and Balunaini, U., (2006) "Determination of ash mixture properties and Construction of test embankment-Part A" Journal of Transportation Research Program, Final Report, FHWA/IN/JTRP-2006/24. Purdue University, W. Lafayette, Indiana.

[15] Robert, F.K. and Kenneth, S.S. (1993) "Trace elements in coal and coal combustion residues" Lewis Publishers.

[16] Goutnam, K. P. and Venkatappa, R., (2008) "Model studies on geosynthetic reinforced double Layer system with pond ash overlain by sand" Indian Institute of Technology Delhi. New Delhi, India.

[17] Das, S.K. and Yudhbir, (2006) "Geotechnical Properties of low calcium and high calcium Fly ash" Journal of Geotechnical and Geological Engineering. Vol. 24 pp. 249-263.

[18] Fawzan, A.A., (2010) "Bottom Ash as a Sand replacement in concrete mix". Unpublished B.Eng. Thesis. KLIUC, Malaysia.

[19] Naganathan, S., Subramaniam, N. & Mustapha, K.N. (2012) "Development of Brick using Thermal Power Plant Bottom Ash and Fly ash". Asian Journal of Civil Engineering (Building and Housing) Vol. 13 No. 1 pp. 275-287.

[20] Awang, A., Marto, A. & Maktar, A.M., (2011) "Geotechnical Properties of Tanjung Bin Coal Ash Mixtures for Backfill Materials in Embankment Construction." EJGE Vol. 16, Bund. L pg. 1515-1531.

[21] Arenas, C.G., Marrero, M., Leiva, C., Solis-Guzman, J., Arenas, L.F.V. (2011) "High fire resistance in blocks containing coal combustion fly ashes and bottom Ash. Waste Management. Vol. 31, Issue 2 pp. 246-252.

[22] Andrade, L.B., Rocha, J.C. and M. Cheriaf., (2007) "Evaluation of concrete incorporating Bottom ash as a natural aggregates replacement". *Waste Manage*, Vol. **27** (9), pp. 1190–1199

[23] Neville, A.M. and Brooks, J.J., (2003) "Concrete Technology" Pearson education limited.

[24] Andrade, L.B., (2004) "Methodology of assessment to use of bottom ash of thermoelectric Power plants as aggregate in concrete". M.Sc. Thesis, Department of Civil Engineering Federal University of Santa Catarina, Brazil [in Portuguese].

[25] Yetgin, S. and Cavder, A., (2006) "Study of effects of natural pozzolan on properties of cement Mortars". *J Mater Civ. Eng*, ASCE. Vol. 18 (6) pp. 813-816.

[26] Kurama, H. and Kaya, M., (2008) "Usage of Coal Combustion bottom ash in concrete mixture" Construction and Building Materials. Vol. 22 pp. 1922-1928.

[27] Kurama, H., Topcu, I.B. and Karakurt, C., (2009) "Properties of the autoclaved aerated concrete Produced from coal bottom ash" Journal of Materials Processing Technology. Vol. 289 Issue 2 pp. 767-773.

[28] Ghafoori, N. and Bucholc, J., (1996) "Investigation of lignite-based bottom ash for structural Concrete". *J Mater Civ. Eng*, Vol. **8** (3), pp. 128–137.

[29] Siddique, R. (2010) "Utilization of coal combustion by-products in sustainable Construction materials. Resources, conservation and recycling. Vol. 54. Issue 12pp. 1060-1066.

[30] Bai, Y., Ibrahim, R. and Basheer, P.A.M., (2010) "Properties of lightweight concrete Manufactured with fly ash, furnace bottom ash and Lytag". International Workshop On Sustainable Development and Concrete Technology. Pp. 77-88.

[31] Andrade, L.B., Rocha, J.C. and M. Cheriaf., (2008) "Influence of coal bottom ash as fine Aggregate on fresh properties of concrete". Construction & Building. 23(2) 609-614

[32] Mandal, S. and Majumdar, D., (2009) "Study on the Alkali activated fly ash mortar" The Open Civil Engineering Journal. Vol. 3 pp. 98-101.

[33] Taylor, G.D., (2000) "Materials in construction, An introduction" Third edition, Longman, U.K.

[34] Wei, L., Naik, T. R. and Golden, D. M., (1994) "Construction Materials Made with Coal Combustion By-Products". Cement, Concrete & Aggregate. American Society For Testing and Materials.

[35] Vilches, L.F. et al., (2007) "Fire resistance characteristics of plates containing a high biomass-Ash proportion". Industrial Engineering Chemical Research. Vol.46 pp. 4825-4829.

[36] NETL (National Energy Technology Laboratory), (2006). "Clean coal technology: Coal utilization by-products. Washington DC; Department of Energy Office of Fossil Energy; Topical report No. 24.

[37] ACAA (2008) "American Coal Ash Association". Coal Combustion Products (CCP) Production & Use. Aurora, CO.

[38] University of Kentucky (2010). "Kentucky Ash".

6

STRENGTH BEHAVIOUR OF BIOMASS FIBER-REINFORCED CONCRETE SLAB

Chai Teck Jung[1], Lee Yee Loon[2], Tang Hing Kwong[3] , Koh Heng Boon[4]

[1,3] Department of Civil Engineering, Politeknik Kuching Sarawak, Km 22, Jalan Matang, 93050 Kuching, Sarawak.
[2, 4] Department of Structure & Material Engineering, Faculty of Civil and Environmental Engineering, Universiti Tun Hussein Onn Malaysia, 86400 Parit Raja, Batu Pahat,Johor.

*Corresponding E-mail : tjchai@poliku.edu.my

ABSTRACT

This paper investigates the compressive strength and flexural strength of biomass fibre-reinforced concrete slab. The main objective of this study is to examine the effect of biomass aggregate and fibre glass on the concrete slab strength. The biomass aggregate is used to replace the natural aggregates. A total of 36 slab samples (250 mm x 600 mm x 50mm thick) and 36 numbers of 150 mm cube samples containing 0%, 30%, 60% and 100% biomass aggregate were prepared. The E-class fibre and Supracoat SP800 were added to increase the strength and to achieve the required workability. All the samples were cured in water with room temperature of around 27°C and tested at the age of 7, 14 and 28 days respectively. The result showed that cube specimens containing 30% biomass aggregate concrete achieved minimum strength of 15 MPa at 28 days. The flexural strength for slab specimens containing 30% biomass aggregate, Supracoat SP 800 and fibre glass gained higher strength compared with control specimens. The 100% biomass aggregate slab achieved 88% of the control specimen strength. The workability was between 150 mm to 170mm slump. The density of the specimens was reduced 20% for cube and 28% for slab compared with control specimens. It can be concluded that the biomass aggregate has good potential as partial aggregate replacement in slab construction when combined with the use of glass fibre and superplasticizer. However, more research needs to be carried out to self-compacting biomass aggregate concrete for sustainable construction.

Keywords: *Biomass aggregate, Compressive strength, Flexural strength, Fibre- reinforced, Self-compacting.*

1.0   INTRODUCTION

Strength of concrete is commonly considered the most valuable property[1]. Strength usually gives an overall picture of the quality of concrete because strength is directly related to the structure of the hydrated cement paste. Moreover, the strength of concrete is almost invariably a vital element of structural design and is specified for compliance purpose. Strength is the most common concrete property tested due to three main reasons[2]. Firstly, the strength of the concrete gives a direct indication of its capacity to resist loads in structural applications. Secondly, strength tests are relatively easy to conduct. Finally, correlations can be developed relating concrete strength to certain other concrete properties that involve much more complicated tests. The two principal reasons for determining the strength of a concrete are to measure the quality or uniformity of the concrete produced and provide a measure of load-carrying capacity in structures[3]. Among concrete strength properties, compressive strength of concrete is one of the most important technical properties. This is because in most structural applications, concrete is employed primarily to resist compressive stresses. In those cases where other stresses are of primary importance, the compressive strength is still frequently used as a measure of the resistance because it is the most convenient way to measure[4]. This is also highlighted in British Code of Practice, BS8110 :1997 [5] where most of the stresses checking are related to compressive strength. For example, BS8110: Part 1, clauses 3.4.5.2 and 3.4.5.8 state that the nominal shear stress of beam must not exceed $0.8f_{cu}^{1/2}$ or 5 N/mm^2 where f_{cu} is the characteristic compressive strength on 28 days.

2.0   LITERATURE REVIEW

Compressive strength is one of the important mechanical properties of concrete. In structural concrete design, the characteristics strength of concrete refers to the 28 days compressive strength [5], [6], [7]. The compressive strength of concrete or mortar is usually determined by submitting a specimen of constant cross section to a uniformly distributed increasing axial compression load in a suitable testing machine until failure occurs [4]. Compressive test is the most common test carried out on concrete. This is because of four reasons [8] : (1) concrete has low tensile strength compared to its compressive strength, thus it is used primarily in compressive mode; (2) it is assumed that most of the important properties of concrete are directly related to the compressive strength; (3) the structural design codes are mainly based on the compressive strength of concrete; and (4) the test is easy and economical. The shape of the specimens for compressive strength testing can be cubes [9] or cylinders [10]. However, the compressive strength obtained from cylinders is lower than cubes specimens [4]. The compressive strength determined by cylinders is around 80% to 96% [6] of the compressive strength determined by cubes. Generally, cube test is widely used at the construction sites in Malaysia [11]. A minimum of six numbers 150 mm concrete cubes are prepared during the concreting work. These cubes are used to determine the compressive strength at 7, 14 and 28 days. Flexural strength or modulus of rapture (MOR) is a mechanical parameter for brittle material's ability to resist deformation under load. The transverse bending test is most frequently employed, in which a beam specimen in rectangular cross section is bent until fracture using a two point flexural test technique [12].

The flexural strength represents the ultimate strength of materials in bending is taken corresponding to the ultimate moment by the elastic relationship[13]. It is measured in terms of stress. The proposed flexural strength where at the time of delivery to the work site, the average flexural strength in the test specimens shall be 4.5 MPa, with no individual unit less than 4 MPa[14]. According to J. Newman and S. C. Ban, (2003), the requirements for lightweight aggregate for structural use are that it has strength sufficient for a reasonable crushing resistance based on BS EN 13055, concrete strengths excess 20 MPa and produces compacted concrete with an oven-dry density in the range 1500–2000 kg/m^3. Lightweight-concrete is defined by BS EN 206-1 as having an oven-dry density of not less than 800 kg/m^3 and not more than 2000 kg/m^3 by replacing dense natural aggregates either wholly or partially with lightweight aggregates. Lightweight aggregate is defined in BS EN 13055 as any aggregate with a particle density of less than 2.0 Mg/m^3 or a dry loose bulk density of less than 1200 kg/m^3. These properties mainly derive from encapsulated pores within the structure of the particles and surface vesicles. For structural concrete suitable aggregates should require a low cement paste content and have low water absorption [15], [16]. For slab construction, the flexural strength must be identified because the design of slab or pavement is based on the theory of flexural strength. Therefore, laboratory mix design based on flexural test may be required. What is flexural strength? Flexural strength is one of the tests to measure the tensile strength of concrete to resist failure in bending such as beam and slab. According to T. H. G. Megson, (2007), the materials ultimate strength in bending is defined by the modulus of rapture (MOR). This is taken to be the maximum direct stress in bending ($\sigma_{x,u}$) which corresponding to the ultimate moment (M_u) and is assumed to be related to ultimate moment by elastic relationship in equation 1[17].

$$\sigma_{x,u} = [M_u/I]\, y_{max} \quad \ldots\ldots\ldots\ldots\ldots\ldots\ldots \text{Equation 1}$$

where, M_u = Moment of resistance at the section
        I   = Second moment of area of the section
        Y_{max} = Distance of the layer under stress from the neutral axis
        σ = Stress

3.0   EXPERIMENTAL PROGRAMME

3.1   MATERIALS PREPARATION

The raw materials for the preparation of concrete mix are ordinary Portland cement, natural fine aggregate, biomass aggregate (fine and coarse) and water. The biomass aggregate used for this research was collected from the plywood factory (Linshanhao Plywood Sarawak Sdn Bhd) located at Demak Laut Industrial Park, Jalan Bako, Kuching. The bulk density of the biomass aggregate was determined for the classification as lightweight aggregate. The biomass aggregate was sieved manually using 5 mm sieve to separate it as fine and 10 mm sieve for coarse aggregate before it was used to replace the natural aggregate in this study. The purpose of replace fine and coarse aggregate in the study will lead to the solution for disposal problem in plywood industry. Chemical compound of biomass aggregate was determined using X-ray fluorescence (XRF). The results are shown in Table 1. The mechanical properties of glass fiber Class-E are shown in Table 2.

Table 1 : Chemical compound of biomass aggregate

Chemical Name	Formula	Concentration
Carbon dioxide	CO_2	0.10%
Silicon dioxide	SiO_2	29.10%
Calcium oxide	CaO	18.40%
Barium oxide	BaO	14.40%
Aluminium oxide	Al_2O_3	13.10%
Sodium oxide	Na_2O	5.88%
Iron oxide	Fe_2O_3	5.75%
Magnesium oxide	MgO	4.78%
Potassium oxide	K_2O	4.38%
Sulfur Trioxide	SO_3	1.28%
Nitrogen Dioxide	P_2O_5	1.23%
Manganese oxide	MnO	0.62%
Titanium oxide	TiO_2	0.57%
Strontium oxide	SrO	0.34%
Vanadium	V	0 < LLD

Table 2 : The Mechanical Properties of Class-E Glass Fiber

Parameters	Values
Tensile Strength (MPa)	3500-3600
Modulus Elastic (GPa)	74-75
Expansion (%)	4.8
Density (kg/m^3)	2600
Perimeter (μm)	8-12

3.2 MIX PROPORTIONS

There is one series of concrete mix for the experimental work in the present study. The concrete mix consists of four types of mix proportion with the replacement of natural fine and coarse aggregate with fine and coarse biomass aggregate ranging from 0%, 30%, 60% and 100%. Class-E Glass fiber with the amount of 1kg glass fiber /400 kg cement was added to identify the effect to the biomass concrete strength. The compressive strength and flexural strength is determined by the same concrete mix. The mix proportions for the present study are tabulated in Table 3.

Table 3: Mix Proportions for Concrete Mix

Mix ID	Mix Proportion							Replacement of Biomass Aggregate	
	Cement (kg/m^3)	Water (kg/m^3)	Natural Fine Aggregate (kg/m^3)	Natural Coarse Aggregate (kg/m^3)	Water / Cement Ratio	Class E-Glass Fiber (kg)	Suprac oat SP800 (ml)	Coarse (kg/m^3)	Fine (kg/m^3)
C 0	400	176	820.8	1003.2	0.44	1	4000	0	0
C 30	400	176	574.56	702.24	0.44	1	4000	246.24	300.96
C 60	400	276	328.32	401.28	0.69	1	4000	492.48	601.92
C 100	400	292	0	0	0.73	1	4000	820.8	1003.2

3.3 SPECIMENS AND TESTS

The 150 mm cubes and slab size is 600 mm x 250 mm x 75 mm were used for the compressive strength and flexural strength study. The specimens were de-mould after 24 hours and then subjected to water-curing at room temperatures varied from 24 to 34°C. Density of each specimen was recorded during testing. Three 150 mm cubes were used for the compressive strength test and three slabs size 600 mm x 250 mm x 75 mm were used for flexural strength at each age according to BS 1881 [9], [12]. Both compressive strength and flexural strength are tested at the age of 7, 14 and 28 days.

4.0 RESULT ANALYSIS AND DISCUSSION

4.1 PROPERTIES OF RAW MATERIALS

The bulk densities of fine and coarse biomass aggregate, natural fine and coarse aggregate and cement are summarized in Table 4.

Table 4: Bulk Densities of Raw Materials

Materials	Loose Bulk Density (kg/m^3)	Compacted Bulk Density (kg/m^3)
Cement	1195	-
Natural Fine Sand	1244	1320
Natural Coarse Aggregate	1458	1665
Fine Timber Ash Aggregate	432	485
Coarse Timber Ash Aggregate	840	975

Based on the bulk density, fine and coarse biomass aggregate have fulfilled the requirement as lightweight aggregate according to BS 3797. The density of the lightweight aggregate must not more than 1200 kg/m^3 for fine aggregate and 1000 kg/m^3 for coarse aggregate [18].

4.2 CONCRETE DENSITY AND SLUMP

The dry densities and slump test of the specimens are shown in Table 5. The replacement of natural aggregate with biomass aggregate had reduced the density of concrete. A weight reduction of 20% for cube specimens and 28% for slab specimens at the age of 28 days was achieved when 30% to100% of natural fine and coarse aggregate was replaced by fine and coarse biomass aggregate. The reduction in weight of the biomass aggregate concrete compared to the control specimen is due to the higher porosity of biomass aggregate with a lower bulk density compared to natural aggregate. The weight reduction of biomass aggregate concrete is significant especially for 100% biomass aggregate concrete with a reduction up to 20 % for cubes and 28 % for slab compared to natural aggregate concrete. The slump for concrete mix were achieved between the range of 150 mm to 170 mm. According to BS EN 206-1: 2000, the slump class is S3 (100 mm – 150 mm) for mix C100 and S4 (160 mm – 210 mm) for mixes C0, C30 and C60. This shown that the concrete workability is high and towards self-compacting biomass concrete purpose.

Table 5: Concrete Density and Slump

Mix ID	Dry density (kg/m^3)			Slump (mm)	Remark
	7 Days	14 Days	28 Days		
C 0	2363	2326	2318	160	Cube Specimens
C 30	2148	2133	2111	165	Cube Specimens
C 60	1990	1950	1895	170	Cube Specimens
C 100	1963	1874	1859	150	Cube Specimens
C 0	2393	2333	2220	160	Slab Specimens
C 30	2260	2260	1967	165	Slab Specimens
C 60	2113	2047	1753	170	Slab Specimens
C 100	1947	1887	1593	150	Slab Specimens

4.3 COMPRESSIVE STRENGTH

The compressive strength developments for cube specimens are shown in Figure 1. The results showed that all biomass aggregate concrete (C30, C60 and C100) achieved lower compressive strength compared to the natural aggregate concrete (C0).

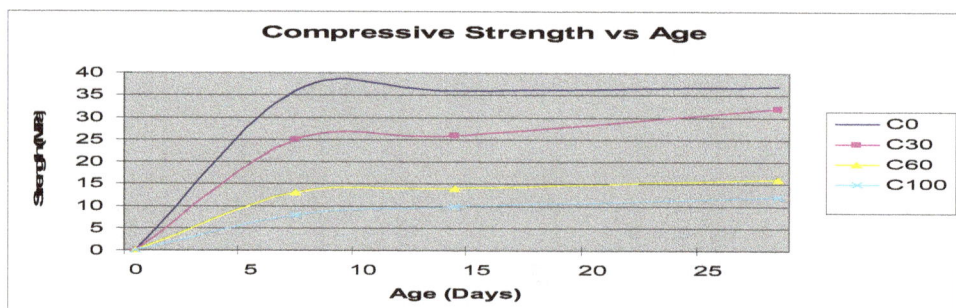

Figure 1 : Compressive Strength for Cube Specimens

Referring to Figure 1, the replacement of 30 % biomass aggregate achieved 32 MPa at the age of 28 days. According to J. Newman and S. C. Ban (2003), the compressive strength development excess 20 MPa can be used for structural purposes. The others mixes C60 and C100 gain a lower compressive strength which achieved 16 MPa and 12 MPa respectively. The strength development for C30 is 3 MPa lower than the targeted one which is 35 MPa. This is maybe due to the mechanical properties of the biomass aggregate even admixture (Supracoat SP 800) was added.

4.4   FLEXURAL STRENGTH

The flexural strength developments for the slab specimens are shown in Figure 2.

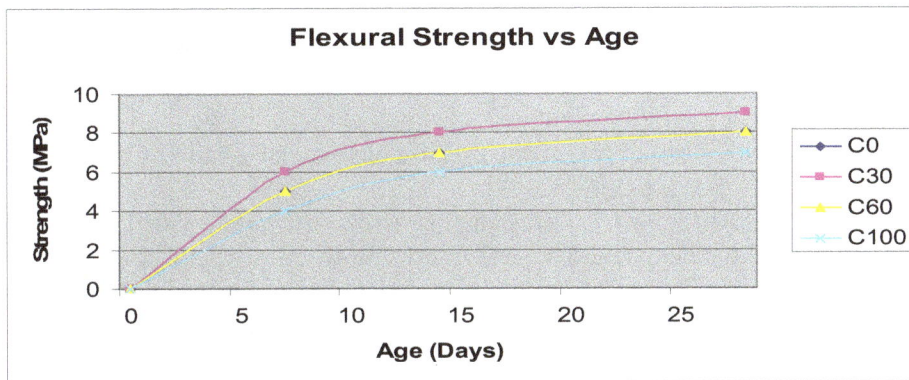

Figure 2 : Flexural Strength for Slab Specimens

Referring to Figure 2, the results showed that biomass aggregate slab with glass fiber produced higher flexural strength (C30) compared with control specimen (C0). The flexural strength for 30 % biomass aggregate with glass fiber is 9 MPa while the control specimen (natural aggregate) without glass fiber and Supracoat SP 800 is 8 MPa at the age of 28 days. The result also showed that C60 have the same flexural strength with control specimen. This result showed that the mixes with fiber glass and Supracoat SP 800 can develope the same strength even though the aggregate is not from the natural resources. The mechanical properties of fiber glass class E also indicate that it behavior is more benefit to flexural strength where the Modulus Elastic is 74 – 75 GPa.

5.0   CONCLUSION

This paper investigated the strength behaviours of biomass aggregate fiber-reinforced concrete slab. Biomass aggregate concrete achieved lower compressive strength may be due to the lower mechanical strength of biomass aggregate. The 30% biomass aggregate concrete were able to achieve the compressive strength of 32 MPa for structural use. According to BS 8110, the strength requirement is 15 MPa which C30 and C60 were fullfilled the British Standard requirement. However, J. Newman and S. C. Ban, (2003) stated that for structural use, the strength must exceed 20 MPa. For flexural strength, the strength development is higher than control specimen for 30 % biomass aggregate added with fiber glass and supracoat SP 800. From the preliminary experimental results obtained in this study, biomass aggregate possesses the potential to be used in construction for materials replacement and willing to save the construction industry cost for one way slab. However, more research needs to be done before the findings can be considered conclusive.

ACKNOWLEDGEMENT

The authors would like to thank the LinShanHao Plywood Sdn Bhd (A Subsidiary Company of WTK Holdings Berhad) and Junda Realty Sdn Bhd for providing the research materials and facilities for this research. The appreciation also express to the Polytechnic of Kuching Sarawak for the financial support and encouragement in research development purpose.

REFERENCES

[1] Neville, A.M. (1995). "Properties of Concrete" 4th Edition. Longman, London.

[2] Klieger, P. & Lamond, J.F. (Editor) (1994). "Significance of Tests and Properties of Concrete and Concrete-Making Materials". ASTM, USA.

[3] Bloem, D.L. (1968). "Concrete Strength in Structures". ACI Journal, Proc. Vol. 65, No.3.

[4] Popovics, S, (1998). "Strength and related properties of concrete – A quantitative approach." New York : John Wiley & Sons.

[5] British Standards Institution (1997). "Structural use of concrete – Part 1: Code of practice for design and construction." London : (BS 8110).

[6] British Standards Institution (2004). "Design of concrete structures – Part 1.1: General rules and rules for buildings." London : (BS EN 1992, Eurocode 2).

[7] B. K. Marsh, (1997). "Design of Normal Concrete Mixes". Second Edition. Construction Research Communications Ltd : United Kingdom.

[8] Mindess, S and J. F. Young (1981). "Concrete." New Jersey : Prentice-Hall.

[9] British Standards Institution (1983). "Testing concrete – Part 116: Method for determination of compressive strength of concrete cubes." London : (BS 1881).

[10] British Standards Institution (1983). "Testing concrete – Part 110: Method for making test cylinders from fresh Concrete." London : (BS 1881).

[11] Koh, H.B., Adnan, S.H. and Lee, Y.L. (2006). "Compressive Strength and Shrinkage of Foamed Concrete Containing Pulverized Fly Ash. Proceeding of Malaysian Science and Technology Conference 2006". Kuala Lumpur : 230-237.

[12] British Standards Institution (1983). "Testing concrete – Part 118: Method for determination of flexural strength." London : (BS 1881).

[13] T. H. G. Megson, (2007). "Structural and Stress Analysis". Second Edition. United Kingdom : Elsevier Ltd.

[14] National Concrete Masonry Association (2002). "Evaluation of Paver Slab Flexural Strength Testing". Research and Development Laboratory : USA

[15] J. Newman and S. C. Ban, (2003). "Advanced Concrete Technology : Constituent Materials". Elsevier Ltd : London.

[16] J. Newman and S. C. Ban, (2003). "Advanced Concrete Technology : Concrete Properties". Elsevier Ltd : London.

[17] T. H. G. Megson, (2007). "Structural and Stress Analysis". Second Edition.United Kingdom : Elsevier Ltd.

[18] British Standards Institution (1990). "Specification for lightweight aggregates for masonry units and structural concrete". London : (BS 3797).

7

THE EFFECTS OF EGGSHELL ASH ON STRENGTH PROPERTIES OF CEMENT-STABILIZED LATERITIC

Okonkwo, U. N[1].; Odiong, I. C[2]. and Akpabio, E. E.[3]

[1,2,3]Department of Civil Engineering, Michael Okpara University of Agriculture Umudike, PMB 7267, Umuahia, Abia State, Nigeria.

*Corresponding E-mail : ugokon2003@yahoo.com

ABSTRACT

Eggshell ash obtained by incinerating Fowls' eggshells to ash has been established to be a good accelerator for cement-bound materials and this would be useful for road construction work at the peak of rainy seasons for reducing setting time of stabilized road pavements. However this should be achieved not at the expense of other vital properties of the stabilized matrix. This is part of the effort in adding value to agricultural materials which probably cause disposal problems. Thus this study aimed at determining the effect of eggshell ash on the strength properties of cement-stabilized lateritic soil. The lateritic soil was classified to be A-6(2) in AASHTO rating system and reddish-brown clayey sand (SC) in the Unified Classification System. Constant cement contents of 6% and 8% were added to the lateritic soil with variations in eggshell ash content of 0% to 10% at 2% intervals. All proportions of cement and eggshell ash contents were measured in percentages by weight of the dry soil. The Compaction test, California Bearing Ratio test, Unconfined Compressive Strength test and Durability test were carried out on the soil-cement eggshell ash mixtures. The increase in eggshell ash content increased the Optimum Moisture Content but reduced the Maximum Dry Density of the soil-cement eggshell ash mixtures. Also the increase in eggshell ash content considerably increased the strength properties of the soil-cement eggshell ash mixtures up to 35% in the average but fell short of the strength requirements except the durability requirement was satisfied.

Keywords: Eggshell ash, Cement, Lateritic soil, Road work, Strength characteristics.

1.0   INTRODUCTION

Lateritic soils are residual soils and are mainly found in the tropical and sub-tropical regions. These are soils formed by the leaching of lighter minerals like silica and in consequent is the enrichment of the heavier minerals like iron and aluminium oxides (sesquioxides). It was stated that the degree of laterization is estimated by silica-sesquioxide ratio [1]. The major part of Nigeria is underlain by basement complex rocks, the weathering of which had produced lateritic materials spread over most of the area. It is virtually impossible to execute any construction work in Nigeria without the use of lateritic soil because they are virtually non-swelling [2].

Soil stabilization techniques for road construction are used in most parts of the world although the circumstances and reasons for resorting to stabilization vary considerably. In industrialized, densely populated countries, the demand for aggregates has come into sharp conflict between agricultural and environmental interests. In the less developed countries and in remote areas the availability of good aggregates of consistent quality at economic prices may be limited. In either case these factors produce an escalation in aggregate costs with maintenance costs. The upgrading by stabilization of materials therefore emerges as an attractive proposition [3]. This will go a long way in actualizing the dreams of the Federal government of Nigeria in scouting for readily cheap construction materials. Internationally, the World Bank equally has been spending substantial amounts of money on research aimed at harnessing industrial waste products for further usage.

Fowls' eggshells are agricultural waste materials generated from chick hatcheries, bakeries; fast food restaurants among others which can litter the environment and consequently constituting environmental problems/pollution which would require proper handling. In the ever increasing efforts to convert waste to wealth, the efficacy of converting eggshells to beneficial use becomes an idea worth embracing. The composition of eggshells indicates that the effect of its ash on cement treated materials should be articulated. It is scientifically known that the eggshell is mainly composed of compounds of calcium which is very similar to that of cement [4, 5, 6]. Eggshell was presented as being composed of 93.70% calcium carbonate, 4.20% organic matter, 1.30% magnesium carbonate, and 0.8% calcium phosphate [7]. Calcium trioxocarbonate (IV) was also presented as an important constituent of egg shells and seashells [8]. Since the dominating compound in egg shell is calcium carbonate, during incineration to ash the calcium carbonate will decompose into calcium oxide and carbon dioxide as shown in Equation 1.

$$CaCO_3 \xrightarrow{heat} CaO + CO_2 \dots\dots\dots\dots\dots\dots\dots\dots\dots\dots\dots\dots\dots\dots\dots(1)$$

Few attempts have been made in the past by researchers with eggshell powder. Common salt was used on an eggshell stabilized lateritic soil with a view of obtaining a good compliment for eggshell powder as a useful stabilizer for road works [10]. Eggshell powder was also used to stabilize lateritic soil for the subgrade of a road work, it was found that eggshell powder possess very low binding property but significantly improved the strength properties of the subgrade soil [11]. Furthermore fly ash, rice husk ash and eggshell powder were used as partial replacement for cement and it was confirmed that the trio when mixed together with cement somewhat has equal strength with that of conventional concrete in mixes [12]. However, very scanty attempts have been made to work on eggshell ash. Eggshell ash has been used as an admixture to cement with a focus on the setting time and it was established as a good accelerator because of extra calcium oxide provided by the addition of eggshell ash [13]. It also satisfied the requirements for initial and final setting times of [14]. The following equations were proposed for the reaction of the hydration process by [13]:

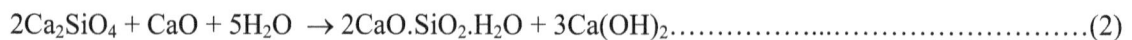

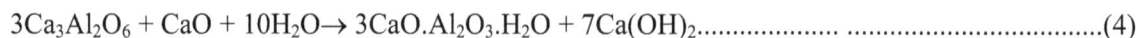

$$2Ca_2SiO_4 + CaO + 5H_2O \rightarrow 2CaO.SiO_2.H_2O + 3Ca(OH)_2\dots\dots\dots\dots\dots\dots\dots\dots\dots\dots(2)$$

$$3Ca_3SiO_5 + CaO + 10H_2O \rightarrow 3CaO.SiO_2.H_2O + 7Ca(OH)_2\dots\dots\dots\dots\dots\dots\dots\dots\dots\dots(3)$$

$$3Ca_3Al_2O_6 + CaO + 10H_2O \rightarrow 3CaO.Al_2O_3.H_2O + 7Ca(OH)_2\dots\dots\dots\dots\dots\dots\dots\dots\dots\dots(4)$$

$$4Ca_4Al_2Fe_2O_{10} + CaO + 17H_2O \rightarrow 4CaO.Al_2O_3.Fe_2O_3.H_2O + 13Ca(OH)_2\dots\dots\dots\dots\dots\dots\dots(5)$$

In the peak of rainy seasons, rainfall frequently interrupts road construction works and it is therefore much desirable to minimize the length of setting time of stabilized matrix as much as possible. The foregoing discovery could likely be a good option for this purpose. Nevertheless, this should be achieved without negatively affecting the other vital properties of the soil-stabilized matrix. Thus, this work focused on investigating into the effects of eggshell ash on the strength properties of cement stabilized lateritic soil.

2.0   METHODOLOGY

The materials used in this study were ordinary portland cement, eggshell ash, lateritic soil and clean tap water. The eggshells were incinerated into ash with temperature of about 500^0C, thoroughly ground and sieved through 75μm sieve in accordance with [14]. The lateritic soil was collected from a deposit in Oboro, Ikwuano Local Government Area of Abia state, Nigeria. It was collected with the method of disturbed sampling technique and air-dried after which preliminary tests were carried out on it in accordance with [15] and [16] for the particle size grading, both contain the details of the tests thus this was to properly classify the lateritic soil. The clay mineral identification was done using the Casagrande's plasticity chart, data in [17]. The summary of the results for the natural lateritic soil are shown in table 1 and the grain size analysis curve shown in figure 1.

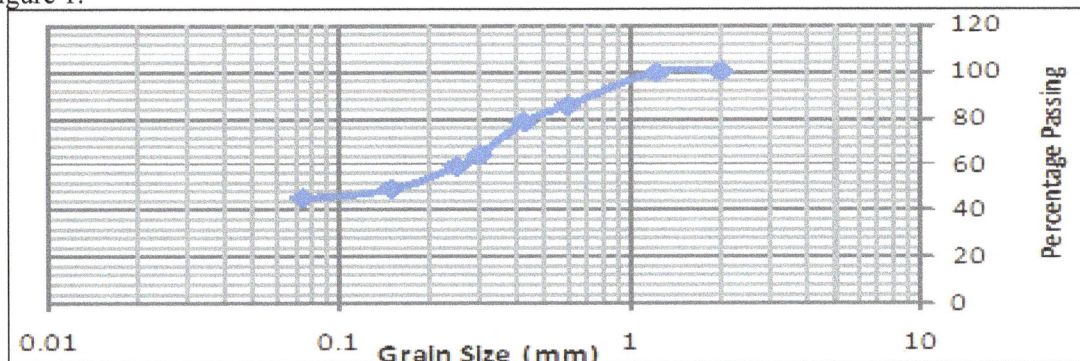

Figure 1: Grading Curve of the Lateritic Soil

The stabilized mixtures were prepared by first of all mixing thoroughly dry pulverized lateritic soil with cement and eggshell ash until a uniform colour was attained after which water was added. Constant cement contents of 6% and 8% with variations of eggshell ash of 0%, 2%, 4%, 6%, 8% and 10% by weight of the dry soil in all proportions of cement and eggshell ash were adopted for the mixes. The amount of water added was determined by the moisture-density relationship in which the Proctor mould was used to place the soil-cement eggshell ash mixtures in 3 layers and 25 blows were given onto each layer. The Unconfined Compressive Strength and California Bearing Ratio specimens were cured with the method of membrane curing. The California Bearing Ratio was modified so as to conform to [18], which stipulates that specimens should be cured for 6 days and immersed in water for 24 hours before testing while the unconfined compressive strength specimens were cured for 7 days before testing. The resistance to loss in strength (durability) was evaluated in accordance with [19] which specifies it as a ratio of the Unconfined Compressive Strength of specimens cured for 7 days and later immersed in water for another 7 days to the unconfined compressive strength of specimens cured for 14 days.

3.0   RESULTS AND DISCUSSIONS

Table 1: Properties of the Lateritic Soil

S/N	Properties	Results
1	Colour	Reddish-brown
2	Percentage passing sieve No 200	45.3
3	Liquid Limit (%)	38
4	Plastic Limit (%)	11
5	Plasticity Index (%)	27
6	Linear Shrinkage (%)	8
7	Specific Gravity	2.75
8	AASHTO Classification [8]	A-6 (2)
9	Unified Classification System	SC (Clayey Sand)
10	Major Clay Mineral Present	Illite (Inorganic Clay of Medium Plasticity)
11	Maximum Dry Density (Mg/m^3)	1.70
12	Optimum Moisture Content (%)	14
13	California Bearing Ratio (%)	9
14	Unconfined Compressive Strength (KN/m^2)	186

The soil was classified to be A-6(2) soil in AASHTO rating system. Though the group is far to the right of the AASHTO table, it is fairly good for road construction works. This is because it has a group index of 2 and also from the point of view of Atterberg limits (liquid limit of 38%, plasticity index of 27% and linear shrinkage of 8%), it is satisfactory. In other words, the clay mineral present in the lateritic soil (illite) is inorganic clay of medium plasticity and the activity is moderately satisfactory thus the soil would be somewhat stable in volume at moisture content variations. However, the high percentage of finer particles (45.3% passing Sieve No 200) in the soil probably caused the soil to have high cement content requirement. It will be also noted that the coarse particles are almost inert in the reaction of the soil with cement rather the finer particles play the major role as the pozzolanic component in the reaction. Therefore higher finer particles results to higher cement content requirement.

Figure 2 shows the relationship between Optimum Moisture Content and eggshell ash content. The Optimum Moisture Content increased progressively from 15% to 17.30% and 16.50% to 18.50% at 6% and 8% cement content respectively with addition of eggshell ash of 0% to 10%. These increments in Optimum Moisture Content with increase in eggshell ash could be attributed to the increased amount of calcium oxide in the mixture as shown in equations (2) through (5) thus this stepped up the rate of hydration reaction which rapidly continued to use up the water in the system. Besides, as the amount of eggshell ash increased; the amount of water required in the system to adequately lubricate all the particles in the soil-cement eggshell ash mixture equally increased. Therefore in the overall, the Optimum Moisture Content continuously increased with increase in eggshell ash content.

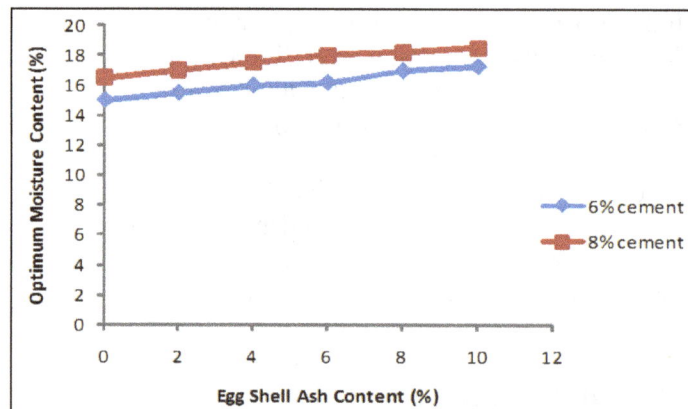

Figure 2: Variations of Optimum Moisture Content with Egg Shell Ash Content

Figure 3 represents the variations in Maximum Dry Density with increase in eggshell ash content. In the corollary of the increase in Optimum Moisture Content with the addition of eggshell ash to the soil-stabilized mixtures was the continuous decrease in Maximum Dry Density from 1.71 Mg/m^3 and 1.74 Mg/m^3 to 1.64 Mg/m^3 and 1.65Mg/m^3 at 6% and 8% cement content respectively with the addition of eggshell ash of 0% to 10%. This could be as a result of the reaction between cement, eggshell ash and fine fractions of the soil as pozzolanic component in which they form clusters like coarse aggregates. These clusters occupied larger spaces thus increasing their volume and consequently decreasing the Maximum Dry Density. Also in some cases the clusters formed were of weak bonding and the disruption was necessary in order to achieve higher level of compaction of the soil-cement eggshell ash mixtures. Therefore during compaction at a given energy level, part of the compactive effort was lost in overcoming the weak bonds which results in reduced density.

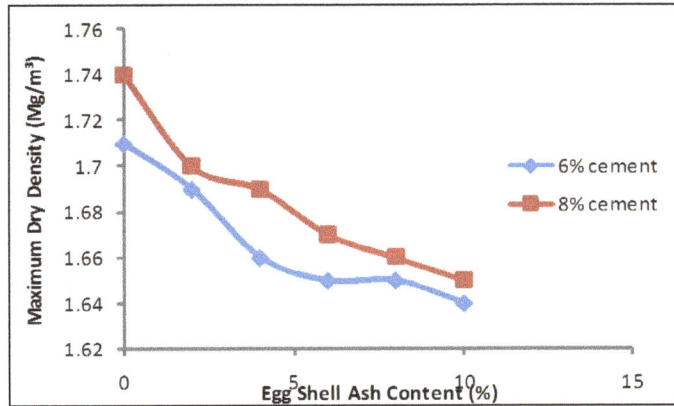

Figure 3: Variations of Maximum Dry Density with Egg Shell Ash Content

Figures 4 and 5 show the variations in California Bearing Ratio and Unconfined Compressive Strength which are the strength properties with increase in eggshell ash content. The California Bearing Ratio (CBR) increased from 26.45% to 56.19% and 82% to 93% at 6% and 8% cement content respectively with the addition of eggshell ash from 0% to 10%. Similarly, the Unconfined Compressive Strength (UCS) at 6% and 8% cement content increased considerably from ($370KN/m^2$ and $471KN/m^2$) for 7 days curing period and ($432KN/m^2$ and $655KN/m^2$) for 14 days curing period to ($614KN/m^2$ and $687KN/m^2$) and ($680KN/m^2$ and $988KN/m^2$) respectively with increase in eggshell ash content from 0% to10%. These could also be linked to the progressive increase in the amount of calcium oxide as a result of the gradual increase in the amount of eggshell ash added to the mixtures. Thus increasing the amount cementitious compounds formed as shown in equations (2) through (5) which equally increased the strength properties.

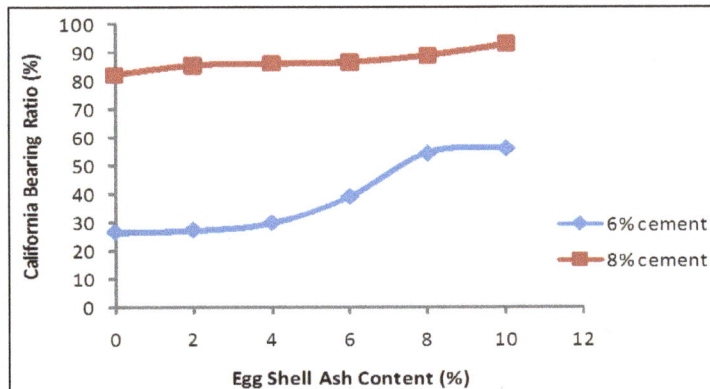

Figure 4: Variations of California Bearing Ratio with Egg Shell Ash Content

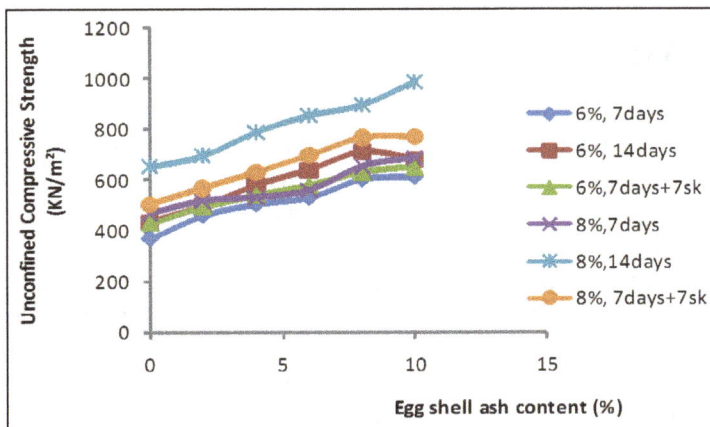

Figure 5: Variations of Unconfined Compressive Strength with Egg Shell Ash

Judging by the 180% value of California Bearing Ratio as stipulated by [18] and the minimum conventional values at 7 days for cement-stabilized soils of $750KN/m^2$ - $1500KN/m^2$, $1500KN/m^2$ - $3000KN/m^2$ and $3000KN/m^2$ – $6000KN/m^2$ for sub-base, base (light trafficked roads) and base (heavily trafficked roads) respectively. The soil-cement eggshell ash mixtures fell short of the requirements however the durability is satisfactory because in most cases of comparing the Unconfined Compressive Strength of 14 days curing period with that of 7 days curing and 7 days soaking, the 20% maximum loss in strength as stipulated by [19] were not exceeded.

The typical products of the hydration process as shown in equations (2) through (5) are composed of the respective hydrates of dicalcium silicates, (2); tricalcium silicates, (3); tricalcium aluminate, (4); tetra calcium aluminoferrite (5) and a common product to the four equations, calcium hydroxide. The calcium silicate hydrates and calcium hydroxide have been described as dominant products of hydration which are produced at the early stage of hydration mainly by the selective hydration of dicalcium silicate and tricalcium silicate [6, 20, 21, 22]. Between the two foregoing, the tricalcium silicate reacts first and dominates the reaction within first few days of hydration [6, 22]. Tricalcium silicates was described as the most important phase of cement and the calcium silicate hydrate gel resulting from this reaction is reported to be principally responsible for the mechanical properties of hydrated cement [20, 22].

Therefore in order to economically improve on the soil material available and also to take care of the huge amounts of calcium hydroxide produced as a result of the hydration reaction of cement and eggshell ash as shown in equations (2) through (5) which might be disadvantageous because it could be a potential source of instability [3], any of the cheap agricultural materials that is rich in silica (pozzolanic material) like bagasse ash, rice husk ash among others should be used together with the eggshell ash and cement. The silica provided by any of these materials will react with the excess amounts of calcium hydroxide produced after hydration reaction of cement and eggshell ash to further produce additional calcium silicate hydrates which is very vital for strength development. The additional amount of calcium silicate hydrates produced will depend on the amount of calcium hydroxide given out from the hydration reaction of cement and eggshell ash. The strength of the resulting stabilized matrix will be tremendously improved considering the amounts of calcium silicate hydrates that would be produced in equations (6) through (8) to meet the requirements of the evaluation standards for strength properties. The following are the proposed equations for the reaction between silica and calcium hydroxide, the common product of hydration reaction of cement and eggshell ash:

$$3Ca(OH)_2 + 3SiO_2 \longrightarrow 3CaO.SiO_2.H_2O \dots\dots\dots\dots\dots\dots\dots\dots\dots\dots\dots(6)$$

$$7Ca(OH)_2 + 7SiO_2 \longrightarrow 7CaO.SiO_2.H_2O \dots\dots\dots\dots\dots\dots\dots\dots\dots\dots\dots(7)$$

$$13Ca(OH)_2 + 13SiO_2 \longrightarrow 13CaO.SiO_2.H_2O \dots\dots\dots\dots\dots\dots\dots\dots\dots\dots(8)$$

4.0   CONCLUSIONS

After the investigation into the effect of eggshell ash on the strength properties of a reddish-brown lateritic soil classified to be A-6(2) in the AASHTO rating and SC (Clayey Sand) in the Unified Classification System, the following conclusions were drawn:

1.   The lateritic soil is fairly good for road construction work.
2.   The increase in eggshell ash content increased the Optimum Moisture Content but reduced the Maximum Dry Density of the soil-cement eggshell ash stabilized lateritic soil.
3.   The increase in eggshell ash content increased the strength properties of the cement-stabilized matrix up to about 35% averagely.

4.   The other strength properties of the soil-cement eggshell ash mixture fell short of the requirements for stabilized materials whereas the durability requirement was satisfied.
5.   On the basis of this investigation, it would be more beneficial to use any other cheap agricultural material that is rich in silica together with cement and eggshell ash in the lateritic soil. However, detailed investigation will be needed to determine the adequate requirement of the preferred agricultural material.

ACKNOWLEDGEMENTS

The authors sincerely wish to express profound gratitude to GEOTECHNICAL ENGINEERING SERVICES LIMITED, Calabar, Nigeria and her technical staff for their support during the laboratory work.

REFERENCES

[1] Makasa, B. (2004); Internet Resource Material, International Institute for Aerospace and Earth Science (ITC) Section, Engineering Geology, Kanaalwey. Netherlands. www.aerospaces.org.ne/ physicalearthscience.

[2] Osinubi, K. J.(1998); "Permeability of Lime Treated Lateritic Soil", Journal of Transportation Engineering, American Society of Civil Engineers, Vol.124, No. 5, pp. 143-152.

[3] Sherwood, P. T. (1993); "Soil Stabilization with Cement and Lime": State-of-the-art Review, Department of Transport, Transport Research Laboratory, United Kingdom.

[4] CCA (1979); Concrete Practice Cement and Concrete Association. Wexham Springs, Slough SL3 6PL.

[5] Shirley, D. C. (1980); Introduction to Concrete, Cement and Concrete Association, Wexham Springs, Slough SL3 6PL.

[6] Neville, A. M. (2003); "Properties of Concrete", Fourth Edition, Second Indian Reprint, Pearson Education, India.

[7] Winton, A. L. (2003); "Poultry Eggs", Agrobios Publishers, Behind Nasrani Cinema, India.

[8] AASHTO (1986); "Standard Specifications for Transportation Materials and Methods of Sampling and Testing 14th Edition", American Association of State Highway and Transportation Officials Washington D. C

[9] Odesina, I.A. (2008); "Essential Chemistry for Senior Secondary Schools", Second Edition, Tonad Publishers Limited, Lagos, Nigeria.

[10] Amu, O. O. and Salami, B.A. (2010); "Effect of Common Salt on Some Engineering Properties of Eggshell Stabilized Lateritic Soil", APRN Journal of Engineering and Applied Science, Asian Research Publishing Network, Vol. 5, No. 9, ISSN 1819-6608.

[11] Olarewaju, A.J., Balogun, M.O. and Akinlolu, S.O. (2011); "Suitability of Eggshell Stabilized Lateritic Soil as Subgrade Material for Road Construction", Electronic Journal of Geotechnial Engineering (EJGE), Vol. 16, pp 889-908.

[12] Jayasankar, R.; Mahindran, N. and Ilangovan, R (2010); "Studies on Concrete Using Fly Ash, Rice Husk Ash and Egg Shell Powder", International Journal of Civil and Structural Engineering, Integrated Publishing Services. ISSN 0976-4399, Vol. 1, No 3, pp 362-372.

[13] Mtallib, M.O.A. and Rabiu, A. (2009); "Effects of Egg Shells Ash on The Setting Time of Cement", Nigerian Journal of Technology, University of Nigeria Nsukka,. ISSN 1115-8443, Vol. 28, No.2, pp. 29-38.

[14] BS 12 (1991); "Specifications of Portland Cement", British Standards Institute, London

[15] BS 1377 (1990); "Methods of Testing Soils For Civil Engineering Purposes", British Standards Institute, London.

[16] ASTM (1993); "Annual Book of ASTM Standards, Soil and Rock: Building Stones: Peats", American Society For Testing Materials, Philadelphia.

[17] Mitchell, J. K. and Kenichi, S. (2005); "Fundamentals of Soil Behavior", John Wiley & Sons, Inc. New Jersey, USA.

[18] Nigeria General Specification (1997); "General Specifications (Roads and Bridges)", Vol.II, Federal Ministry Works and Housing, Abuja, Nigeria.

[19] BS 1924 (1990); "Methods of Test For Stabilized Materials For Civil Engineering Purposes", British Standards Institute, London.

[20] Escalante-Garcia, J.I. and Sharp, J.H. (2004); "The Chemical Composition and Microstructure of Hydration Products in Blended Cements, Cement and Concrete Composites", Elsevier Limited., Vol.26, Issue 8, pp. 967-976.

[21] Kjellsen, K.O. and Justnes, H. (2004); "Revisiting the Microstructure of Hydrated Tricalcium Silicate- a Comparison to Portland Cement. Cement and Concrete Composites". Elsevier Limited. Vol, 26, Issue 8, pp. 947-956.

[22] Scrivener, K.L. (2004); "Backscattered Electron Images of Cementitious Microstructures: Understanding and Quantification, Cement and Concrete Composites", Elsevier limited, Vol. 26, Issue 8, pp 935-945.

8

UTILIZATION OF EUCALYPTUS OIL REFINERIES WASTE FOR CEMENT PARTICLE BOARD

Rudi Setiadji* and Andriati Amir Husin

Experimental Station of Building Materials, Research Institute for Human Settlements

*Corresponding E-mail : kolaka_80@yahoo.com

ABSTRACT

Utilization of eucalyptus oil refinery waste in the manufacture of building material component of cement particle board is expected to reduce the price of housing units. This research used laboratory experimental methods, eucalyptus oil waste in the form of branches an twigs from eucalyptus tree. The variation of the testing were mixtures composition of the particle: cement, additives as accelerators, cold press load during manufacture of cement particle board. Cold press duration of cement board was 24 hours. The sizes of particle boards were (40 x 40) cm^2 and 13 mm thick. The samples were tested for its density, water content, water absorption, flexural strength, thickness swelling, adhesion strength, and the nails pull out strength.

Keywords: Particle, Eucalyptus, Cement board

1.0   INTRODUCTION

Prices of building materials can be reduced by use of basic materials derived from waste. Utilization of waste in this study was eucalyptus oil refinery waste in the form of eucalyptus twigs. Eucalyptus plant in Forest Stakeholder Unit (KPH) Indramayu were scattered at five sections of Forest Stakeholders Unit (BKPH). Based on existing data (KPH Perhutani Indramayu, 1999) sustaining areas for eucalyptus oil refineries were amounted to 5,300 hectares with an estimated production of eucalyptus leaves (DKP) at 3 tons/ha. The utilization for eucalyptus oil processing plant is 60 tons/day. Eucalyptus oil refinery waste form might be liquid, leaves, and eucalyptus wood. Details of the waste generated were 40% leaves for fuel, and 60% twigs with a diameter of 0.5 to 1 cm of 550 kg/ton dry. Means that in 60 ton/day oil refinery waste in the form of wood was 550 kg x 60 tons/day equivalent to 33,000 kg. Utilization of eucalyptus oil refinery waste as alternative raw materials for cement particle board is expected to reduce the price of housing units.

Figure 1: Landfill of eucalyptus oil refinery waste

Based on data from KPH Perhutani Indramayu, there are some eucalyptus oil factory in Indonesia which is under Perhutani and five are located in the island of Java, Sukabumi, Kuningan, Indramayu, Gundih, and Ponorogo. As example of the eucalyptus oil factory (PMKP) Jatimunggul in Indramayu has a production target which increases every year, 12,000 tons/year of eucalyptus oil were processed in 2008 and this factory was included as the largest in Southeast Asia. At present, all of the waste is used as a plantation compost. Eucalyptus oil produced from every 1 ton of material is equivalent to 7 kg. Most of the waste, approximately 6,000 tons/year, can be used for boiler fuel so that the average remaining 6,000 tons/year of waste that has not been exploited economically.

Based on previous research and available potential resources, there is a need for research about properties of cement bonded particle board with eucalyptus oil refinery waste and variation of chemicals additive that complied with standard requirements.

2.0   LITERATURE REVIEW

Many previous research on cement bonded particle board has been conducted. Aini [1] studied cement board with oil palm stem, the flexural strength only up to 14,36 kg/cm^2. It was predicted because of insufficient clamp load. Ashori et al. [2] described that there is a positive correlation between the compression ratio of wood materials and their bending strength. Alhedy et al. [3] studied the effect of water soaking periods of time, pressure levels, and cement/bamboo ratio. The effect of the three studied factors was dependent on each other for MOR and hardness. The results showed that all specimens met the specification of the British standard (BS 1105; 1972) for strength properties and the density of the product increased with increasing cement/bamboo ratio. Soaking in water for more than 7 or 10 days was either not significantly different from 3 days or had lower strength and density values; hence, no need for soaking in water for more than 3 days. Water absorption and dimension swelling decreased significantly with increasing cement/bamboo ratio. Also increasing of pressure from 0.05 kg/cm^2 to 0.15 kg/cm^2 did not improve the mechanical and physical properties, this may be due to the low pressure levels used in the study.Aggarwal [4] observed that when the casting of bagasse-cement samples was increased from 1 to 3 N/mm^2, there was about 32% increase in density (1.21 – 1.60 g/cm^3) of the consolidated mass while 7.5% increase ws observed when casting pressure was increased from 3 – 5 N/mm^2 for the same volume of baggase-cement mix. The water absorption of the samples is from 19.5 to 13.4% and 13.4 to 11.6% when the casting pressure is increased from 1 to 3 N/mm^2 and 3 to 5 N/mm^2, respectively. The optimum casting pressure required for bagasse-cement samples is in the range of 2-3 N/mm^2.

Antonios N. P. [5] made cement bonded particleboards with hornbeam (Carpinus betulus L.) wood particles and $CaCl_2$ additive as 3%. Heat hydration test was done to gain the inhibitory index for characterizing the relationship between wood and cement. Results show that hornbeam-cement mixture can be classified as moderate inhibition. Wood : cement ratio used were 1 : 3 and 1 : 4. Increasing wood : cement ratio improved the characteristics of cement bonded particleboards except the modulus of rupture. Idris, A.A [6] studied cement bonded particleboards with reed fiber. Flexural strengths with lime immersion were in range of 40,18 – 70,66 kg/cm^2, while flexural strengths with water immersion were in range of 28,25 – 61,24 kg/cm^2.

Ling Fei Ma [7] concluded that there is a relationship between cement hydration with mechanical properties of cement board from wood and other lignocellulosic materials. Mechanical performances can be predicted from total of energy released from hydration proccess. Materials being used was wood chip of sugi, hinoki, kenaf, bamboo, rice hull, and rice straw. Cement portland type I has added with $MgCl_2$, Na_2CO_3, $NaHCO_3$, $NaSiO_3$, $CaCl_2$ additive as 0%, 2,5%, 5%, 10%, and 15%. Increasing additive content increased the modulus of elasticity and modulus of rupture with maximum additive content range between 5% to 10%.

Sudin and Swamy [8] made bamboo flakes and oil palm fibres, tested the sugar content , and its effect on strength development of cement Portland mixture. Accelerator materials used were $CaCl_2$, $MgCl_2$, $Al_2(SO_4)_3$, $Al_2(SO_4)_3+Na_2SiO_3$ as 2%, fly ash, rice husk ash, and latex for cement partial replacement as 10%, 20%, and 30%. Bamboo-cement board with bamboo : cement ratio as 1 : 2,75 with aluminium sulphate additive as 2% complied with local standard MS 934 requirement. Oil palm fibre-cement board with cement replacement range between 10% to 20% satisfied the local standard MS 934 requirement.

3.0   METHODOLOGY

The research using laboratory experimental methods carried out in the Building Materials Department in Research Institute of Human Settlements of the Ministry of Public Works in Bandung. Waste eucalyptus branches and twigs were obtained from PMKP Jatimunggul Indramayu. Eucalyptus oil waste was tested for heat of hydration. The composition of the mixture between the particles : cement by mass were 1 : 3, 1 : 4 and 1 : 5. Additional materials as an accelerator were $CaCl_2$, $MgCl_2$, $Al_2(SO4)_3$. Cold press loads during the process of cement particle board making were 20 kg/cm^2, 25 kg/cm^2, 40 kg/cm^2 and 50 kg/cm^2. Cold press duration of cement board was 24 hours. The size of particle board was (40 x 40) cm^2 and 13 mm thick. The number of repetitions of each test were 3 times and were tested for density, water content, water absorption, flexural strength, thickness swelling, strong adhesion, and the nails pull out strength. The test results were compared with SNI 03-6861.1-2002 Building Material Specifications Part A (Non Metallic Building Materials) [9] and ISO 8335-1987 Standards of Cement-bonded Particleboard [10]. Eucalyptus wastes was crushed resulting of 4.8 mm maximum size. Physical properties of eucalyptus particles are given in Table 1

Table 1: Physical properties of eucalyptus particles

Properties	Results	
Specific gravity	0.21	
Water absorption	9.97	%
Water content	1.70	%

4.0   RESULTS AND DISCUSSIONS

4.1   HEAT OF HYDRATION

Inhibitory index of neat cement in Table 2 is lower than the index of eucalyptus particles–cement mixture, it shows that eucalyptus particles has wood-extractives that reduced the maximum hydration temperature. Time to maximum temperature of eucalyptus particles–cement mixture was reduced compared to the neat cement, contrary to a previous study by Ashori et al [2]. Particles size might have possible effect of wood-extractive, as finer particles expose more surface area to the cement paste and thus more extractives can enter into solution. Hydration test can be related to mechanical strength as research by Huceng Qi [11] on chromated copper arsenate (CCA)-treated wood-cement composite that with the increase of wood content, the maximum hydration temperature was decreased, and also the splitting tensile strength. But due to the effect of $CaCl_2$, time to reach maximum hydration temperature for CCA-treated wood cement composite was even shorter than that of neat cement, same behaviour with the results in Table 2 from this eucalyptus-cement mixture study. Same result by Wei et al. [12] that there was a positive trend between hydration characteristic and strength of wood cement-baced composite. The higher the Tmax, the larger were the strength values of Modulus of Rupture (MOR) and Internal Bonding (IB).

The hydration test was carried out primarily to determine the effect of the accelerator in this study on the early hydration behaviour of cement. As given in Table 2 that mixture composition without accelerator have greater value of inhibitory index than mixture composition of neat cement, it shows that cement is not well hydrated eventough the cement content was increased up to composition of cement to eucalyptus particles as 5 : 1 and the inhibitory index cannot be increased.

Accelerator addition as 2% is increased the inhibitory index, probably due to the capacity of the accelerator minimize the adverse effect of the inhibitory chemicals released from wood and also to accelerate the cement hardening and setting similar with experiment by Antonios [5]. The ideal material should provide higher maximum temperature and lower time to achieve the maximum temperature. Best performance for eucalyptus to cement composition is resulted from $MgCl_2$ followed by $CaCl_2$ addition. Maximum hydration time is reached faster with $CaCl_2$ addition. Addition of $Al_2(SO_4)_3$ is not recommended for eucalyptus particles to cement composition. Based on smallest inhibitory index and hydration time, all samples was made with composition of eucalyptus particle to cement as 4 : 1 with $CaCl_2$ addition as 2%.

Table 2: Heat of hydration test results

Eucalyptus particles : cement	Accelerator	t [hours]	T [°C]	S	I
0 : 1	-	8.0	28.2	0.27	0.000
1 : 3	-	6.5	27.3	0.10	0.893
1 : 4	-	6.5	27.3	0.10	0.893
1 : 5	-	6.5	27.3	0.10	0.893
1 : 4	$CaCl_2$	5.0	28.9	0.13	- 0.140
1 : 4	$MgCl_2$	7.0	29.0	0.20	- 0.409
1 : 4	$Al_2(SO_4)_3$	7.0	28.0	- 0.09	0.519

4.2   DENSITY OF CEMENT PARTICLES BOARD

Eucalyptus-cement board density ranged between $1200 - 1310$ kg/m^3, increasing of clamp load increased densities as given in Figure 2 due to the thickness is not limited by stopbar as the packing increased and volume decreased but the weight is relatively constant. Same result by Aggarwal [4] that the increase in casting pressure results in the formation of more compacted fibre-cement mass, thereby decreasing void volumes within samples and hence the thickness of the consolidated mass. In the case of bagasse-cement samples there was a continuous increase in density with the increase in casting pressure.

Density test results are given in Table 3. Samples density with stopbar treatment, clamp load as 20 kg/cm^2 and 25 kg/cm^2 are complied with SNI 03-6861.1-2002 requirement [9]. All samples are satisfied with the ISO 8335-1987 requirement [10]. There is a correlation between density and mechanical properties due to enhance wood densification, elimination of gaps, and improved connection between cement matrix and particles. Density, water absorption, and porosity are all interrelated physical properties. When particle content is increased, water absorption increases, whereas density decreases as seen in comparison of Figure 2 and Figure 3.

Clamping was done by Universal Testing Machine (UTM) for 24 hours and cured at room temperature. All samples thickness with stopbar greater than designed thickness as 13 mm, it was predicted because of less clamping duration so samples swelling back. Board thickness with clamp load as 25 kg/cm^2 is same with designed thickness so it can be used as the maximum clamp load of board without using 13 mm stopbar.

4.3   WATER ABSORPTION

Eucalyptus-cement board properties of water absorption ranged between $22 - 28\%$, increasing clamp load decreased the water content as given in Figure 3. High pressure from the clamp minimize voids and reduce water infiltration. Wood particles have greatest absorption among board materials when exposed to water but clamp load may play dominant role in reducing this effect owing to more cement matrix coverage by pressure, leading to blocking the flow of water into wood particles. Higher clamp load results higher density and increasing density decreased water absorption same results by Aggarwal [4]. There is no requirement for maximum water absorption in SNI 03-6861.1-2002 [9] or ISO 8335-1987 [10].

4.4   WATER CONTENT

Eucalyptus-cement board water content ranged between 11.8 – 14.5%, increasing clamp load decreased water contents as given in Figure 4. Samples water content with clamp load are complied with ISO 8335-1987 maximum requirement as 12% [10] due to clampping pressure forced the water leak out. Cement particle board making with clamp load treatment results lower water content than cement particle board making with stopbar.

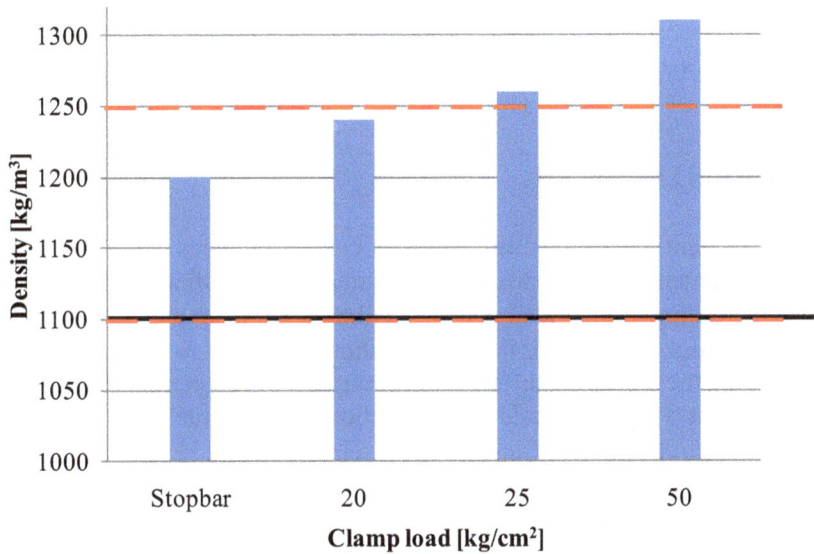

Figure 2: Density of cement-bonded eucalyptus particle board

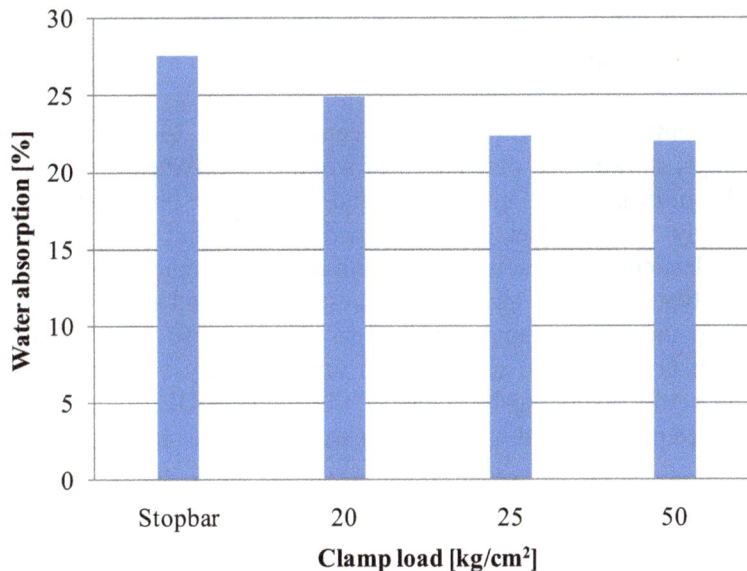

Figure 3: Water absorption of cement-bonded eucalyptus particle board

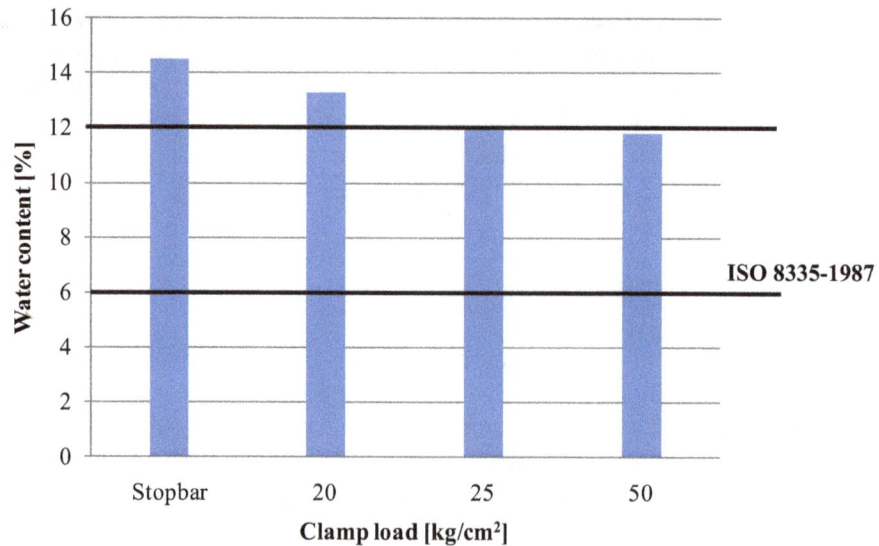

Figure 4: Water content of cement-bonded eucalyptus particle board

4.5   FLEXURAL STRENGTH

Eucalyptus-cement board flexural strength ranged between 22.84 – 47.82 N/mm^2 as showed in Figure 5. Cement bonded particle board samples is satisfied with SNI 03-6861.1-2002 requirement [9]. Increasing clamp load increased the flexural strength as given in Figure 5, probably bonds between samples materials composition increased as the clamp load increasing. Due to the pressure of the clamp load, frictional force are developed between particles and the cement matrix. When the system is loaded these frictional forces transmit stresses if the particles are embedded in the matrix over a sufficient length. Huceng Qi [11] stated that particle size has been shown to be a main effect on the strength properties. Non-ideal particle size such as thicker than the practical requirement and the fine particles which were not screened and discarded before the board manufacture would resulted low flexural strength. Eucalyptus particle maximum size used in this study was 4.8 mm and no fine particles discarded, so probably the size and gradation of particles was in ideal range and contributed to the flexural strength.

4.6   MODULUS OF ELASTICITY (MOE)

Eucalyptus-cement board MOE ranged between 1893.50 – 2467.00 N/mm^2 as showed in Figure 6. Ashori et al.[2] showed that percentage of wood-wool mixture, cement, and CaCl$_2$ and their interactions had significant effect on the MOR and MOE of the boards. Those three variabels were fixed in this study so the effect of clamp load can be studied. Increasing clamp load increased modulus of elasticity as given in Figure 6 but none of samples complied with standards. It is predicted that eucalyptus particles have low stiffness or low bond strength with cement. Other reason probably due to increasing clamp load increased density so modulus of elasticity also increased. Similar with Oyagade [13] resulted that board density was linearly and positively related to bending properties including Modulus of Elasticity (MOE) and Modulus of Rupture (MOR) of cement bonded particleboard when cement/wood ratio was held constant.

4.7   THICKNESS SWELLING

Thickness swelling of cement-bonded eucalyptus particle board after soaked in water for 24 hours ranged from 0.13 to 0.38 % are complied with ISO 8335-1987 maximum requirement as

2% [10]. Semple and Evans [14] argue that thickness swelling is highly dependent on particle geometry. It increases with increasing particle thickness and decreasing particle length. The use of thicker particles results in greater heterogeneity and more irregular open board surface which is more easily penetrated by water. Rough surface and greater internal void space caused by the use of thicker particles are responsible for higher thickness swelling. Meneeis et al. [15] also mentioned that by using low cement-wood ratios, the wood particles are not encapsulated by cement, which results in low bonding and therefore in low internal bond values and increased thickness swelling. Based on those previous results, it is predicted that thin and short particle, enough cement-wood ratio, and high clamp load in this study increased the bonding strength and reduced the thickness swelling significantly.

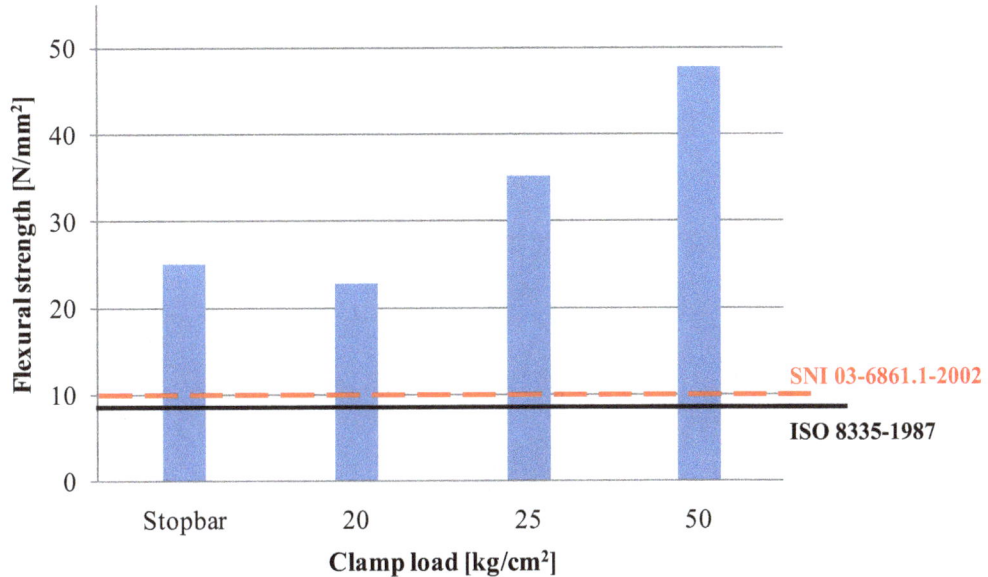

Figure 5: Flexural strength of cement-bonded eucalyptus particle board

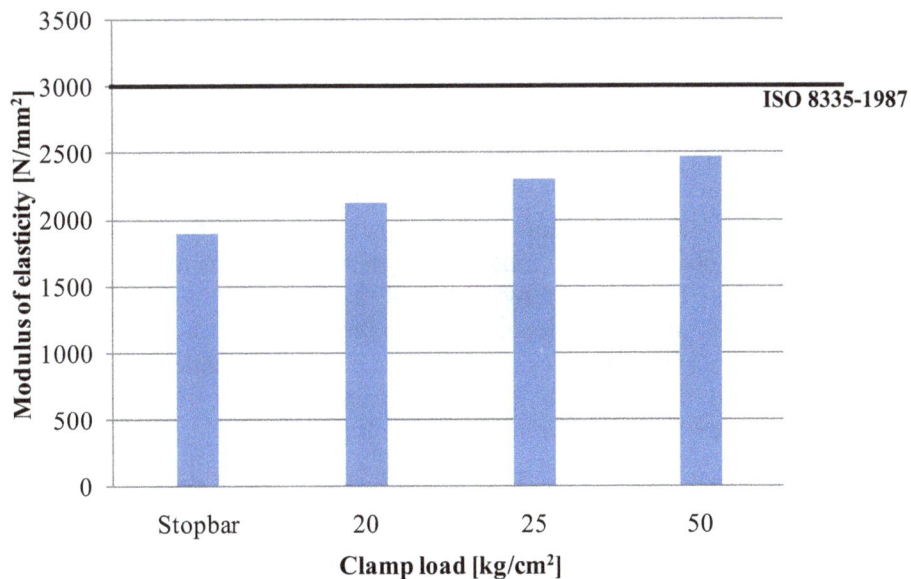

Figure 6: Modulus of elasticity of cement-bonded eucalyptus particle board

4.8   BOND STRENGTH

Eucalyptus-cement board bond strength ranged between $1.21 - 4.37$ kg/cm^2 as showed in Figure 7. Bond strength can be affected by several factors. Huceng Qi [11] stated that bond strength reflects directly the bonding ability of portland cement, the quality of cement hydration, and the compatibility of CCA-treated wood with cement. Higher water/cement ratio contributed to higher bonding strength due to the better hydration of cement and improvement to the fine structure of hydrated cement product. Bond strength of cement-bonded eucalyptus particle board in this study that satisfied with SNI 03-6861.1-2002 requirement [9] only for samples with clamp load as 50 kg/cm^2 as given in Figure 8. It is assumed clamp load also affect the bond strength and clamp load as 50 kg/cm^2 would be the minimum value of clamp load for complied board.

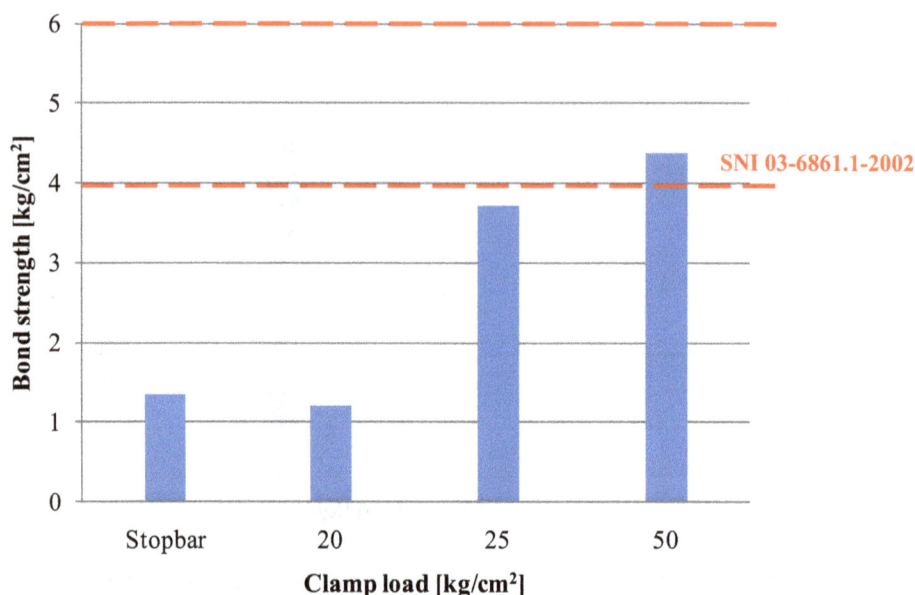

Figure 7: Bond strength of cement-bonded eucalyptus particle board

4.9   NAIL PULL OUT STRENGTH

It is explained in the Wood Handbook [16] that the strength and stability of any structure depend heavily on the fastenings that holds its parts together. Nails are the most common mechanical fastenings used in wood construction to resist witdrawal loads, lateral loads, or a combination of the two which affected by factors of the wood, the nail, and the condition of use. In general, any variation in these factor has a more pronounced effect on withdrawal resistance than on lateral resistance. The resistance of nail to direct withdrawal depends on the density of the wood product, the nail diameter, and depth of penetration. The nail diameter and depth of penetration in this study was conditioned to be uniform, so the density of the cement board defined. Nail pull out strength of of cement-bonded eucalyptus particle boards ranged between $45.00 - 60.33$ kg as showed in Figure 8 and complied with SNI 03-6861.1-2002 requirement [9]. Increasing clamp load increased nail pull out strength as given in Figure 9, probably clamp load increased the density and improve frictions between eucalyptus cement board to nail.

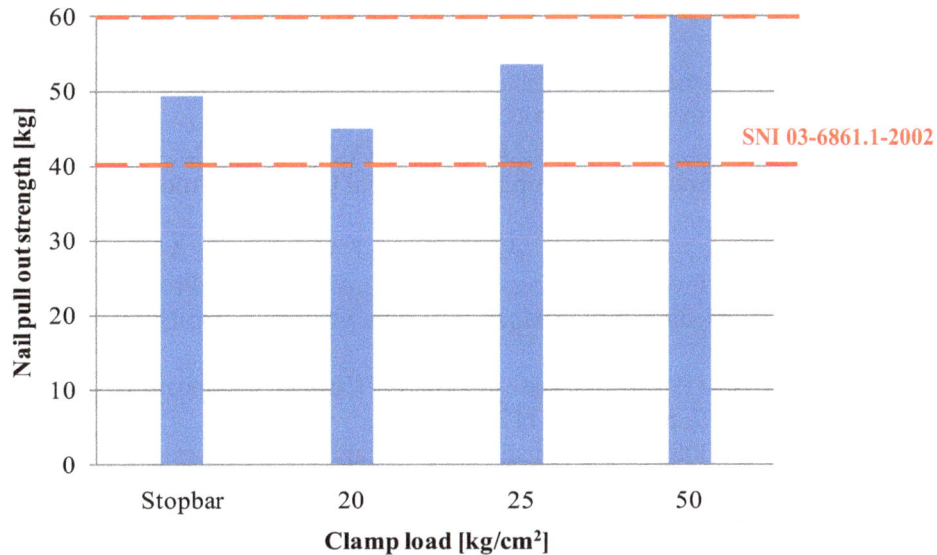

Figure 8: Nail pull out strength of cement-bonded eucalyptus particle board

5.0   CONCLUSIONS

Potential waste eucalyptus oil based on data from KPH Indramayu Perhutani are quite numerous and scattered in various locations in Indonesia. Material in the form of branches and twigs fairly easily obtained in terms of collecting, cleaning, and transportation. The heat of hydration test results showed that although the amount of cement was increased to the composition of eucalyptus fiber: cement at 1 : 5, it could not increase the inhibitory index. Inhibitory index was best for the composition of white cement-eucalyptus fiber using accelerator $MgCl_2$, and the maximum heat of hydration time was reached quickly by using $CaCl_2$, whereas $Al_2(SO4)_3$ results was not optimal for the composition of the cement-eucalyptus fiber. Based on its density, samples was complied with SNI 03-6861.1-2002 except samples with cold press loads at 25 kg/cm^2 and 50 kg/cm^2. The value of water contents that meet the maximum 12% requirement was obtained with a cold press load at 50 kg/cm^2. Water absorptions were ranged between (22.0 to 27.6)% and the lowest value was obtained by cold press load at 50 kg/cm^2. The entire value of flexural strength specimens were complied with SNI 03-6861.1-2002 but the modulus of elasticity were not complied. Value of thickness swelling of the entire specimen thickness were ranged in (0.13 to 0.43)% and were qualified as the maximum value at 2%. Cold press load value at 50 kg/cm^2 could be the minimum boundary for eucalyptus cement board in order to get a strong adhesive that meets the requirements. Nails pull out strength were range in (45.00 to 60.33) kg, complied with SNI 03-6861.1-2002. Based on test results, it was concluded that all specimens of eucalyptus cement board in this study using a mixture of eucalyptus fiber : cement at 1 : 4 and additional materials as much as 2% $CaCl_2$ addition, resulted the best characteristics when combined with cold press load at 50 kg/cm^2.

ACKNOWLEDGEMENT

A gratitude and sincere appreciation are presented for the Research Institute of Human Settlements in Bandung, KPH Perhutani Indramayu, eucalyptus oil factory (PMKP) Jatimunggul in Indramayu for finance, materials, and test facilities support that have been provided.

REFERENCES

[1] Aini, N (2002), "Pemanfaatan Limbah Sawit untuk Bahan/komponen Bangunan Perumahan", Research Institute for Human Settlements, Bandung

[2] Ashori, A., Tabarsa, T., Azizi, K., and Mirzabeygi, R. (2011), "Wood-wool Cement Board Using Mixture of Eucalypt and Poplar", Industrial Crops and Product, Vol 34, pages : 1146 – 1149

[3] Alhedy, A. M. A., Algadir, A. A. Y. A, and Mohamoud, A. E. A (2005), "Effect of Pretreatment and Pressure on Properties of Cement-bonded Products from Oxytenanthera abyssinica", Proceedings of the Meetings of the National Crop Husbandry Committee 40th, pages :.237-244

[4] Aggarwal, L. K. (1995), "Bagasse-reinforced Cement Composites", Cement & Concrete Composites, Vol 17, pages : 107-112

[5] Antonios N. P. (2008), "Mechanical Properties and Decay Resistance of Hornbeam Cement Bonded Particleboards", Research Letters in Materials Science, Vol 2008, Article ID 379749

[6] Idris, A.A. (1994), "Penelitian Pemanfaatan Alang-alang sebagai Papan Semen", Jurnal Penelitian Permukiman,Vol 10, pages : 3-4, Research Institute for Human Settlements, Bandung

[7] Ling Fei Ma (2005), "Manufacture of Bamboo-Cement Particleboard", Institute of Wood Science and Technology, Zhejiang Forestry University, Linan, Zhejiang, China

[8] Sudin, R., and Swamy (2006), "Bamboo and Wood Fibre Cement Composite for Sustainable Infrastructure Regeneration", Journal of Materials Science, Vol 41, pages : 6917-6924

[9] SNI 03-6861.1-2002, "Building Material Specifications Part A (Non Metallic Building Materials)", National Standardization Agency

[10] ISO 8335-1987, "Standards of Cement-bonded Particleboard", International Organization for Standardization

[11] Huceng Qi (2001), "Leaching, Hydration and Physical-mechanical Properties of Spent Chromated Copper Arsenate (CCA)-treated Wood-cement Composite", Master Thesis, Graduate Department of Forestry, University of Toronto

[12] Wei, Y. M., Zhou, Y. G., and Tomita, B. (2000), "Study of Hydration Behavior of Wood Cement-based Composite II : Effect of Chemical Additives on the Hydration Characteristics and Strength of Wood-cement Composites", J Wood Sci, Vol 46, pages : 444-451

[13] Oyagade, A. O. (1989), "Effect of Cement/wood Ratio on the Relationship Between Cement Bonded Particleboard Density and Bending Properties", Journal of Tropical Forest Science, Vol 2(2), pages : 211-219

[14] Semple, K. E, and Evans, P. D. (2007), "Manufacture of Wood-cement Composite from Acacia Mangium. Part II. Use of Accelerators in the Manufacture of Wood-wool Cement Boards from A. Mangium", Wood Fibre Sci, Vol 39, pages : 120-131

[15] Meneeis, C. H. S., Castro, V. G., and Souza, M. R. (2007), "Production and Properties of a Medium Density Wood-cement Boards Produced with Oriented Strands and Silica Fume", Maderas : Ciencia y tecnologia, Vol 9(2), pages : 105-116

[16] Risbrudt. C. D, Ritter. M. A., and Wegner. T. H. (2010), Wood Handbook-Wood as an Engineering Material, Forest Product Laboratory, Forest Service, U.S. Department of Agriculture, Madison.

9

FUZZY MCDM MODEL FOR RISK FACTOR SELECTION IN CONSTRUCTION PROJECTS

Pejman Rezakhani

Department of Architecture and Civil Engineering, Kyungpook National University, Korea

*Corresponding E-mail : rezakhani@knu.ac.kr

ABSTRACT

Risk factor selection is an important step in a successful risk management plan. There are many risk factors in a construction project and by an effective and systematic risk selection process the most critical risks can be distinguished to have more attention. In this paper through a comprehensive literature survey, most significant risk factors in a construction project are classified in a hierarchical structure. For an effective risk factor selection, a modified rational multi criteria decision making model (MCDM) is developed. This model is a consensus rule based model and has the optimization property of rational models. By applying fuzzy logic to this model, uncertainty factors in group decision making such as experts` influence weights, their preference and judgment for risk selection criteria will be assessed. Also an intelligent checking process to check the logical consistency of experts` preferences will be implemented during the decision making process. The solution inferred from this method is in the highest degree of acceptance of group members. Also consistency of individual preferences is checked by some inference rules. This is an efficient and effective approach to prioritize and select risks based on decisions made by group of experts in construction projects. The applicability of presented method is assessed through a case study.

Keywords: *Multi criteria decision making; Risk management; Fuzzy set; Construction management*

1.0   INTRODUCTION

There are many risk factors in a construction project. These risk factors vary from project to project depending on different conditions of a project. The first step to have an effective risk management plan is risk classification. Risk classification is an important step in the risk assessment process, as it attempts to structure the diverse risks that may affect a project. In this study through a comprehensive literature survey of different risk classification approaches, most effective risk factors in a construction project are classified by their source and effect on project objective. Although this classification is comprehensive but it is not restricted and depending on different situations of a project, some new factors can be added to this classification. To make the risk management plan as effective as possible, the most effective risk factors on project objectives should be prioritized and selected through group decision making. Group members consist of different experts in construction industry with variety in experience, knowledge and expertise. In this research we proposed a fuzzy multi- criteria group decision making solution which is based on the Hybrid Rational- Political model. The proposed model has ten steps within three stages. The rest of the paper is organized as follows. In the next section, a literature survey on different methods of risk classification with focus on construction project risks is introduced. This section ends with a suggested hierarchical risk factor classification in a construction project. Then in the subsequent section, the proposed methodology for risk factor prioritization and selection in defined. Applicability of proposed model is assessed through a case study in next section and final section concludes the article.

2.0   RISK CLASSIFICATION

PMBok Version 2008 [1] defines risk classification as a provider of a structure that ensures a comprehensive process of systematically identifying risks to a consistent level of detail and contributes to the effectiveness and quality of the identify risks process. Risk classification is an important step in the risk assessment process, as it attempts to structure the diverse risks that may affect a project. There are many approaches in literature for construction risk classification. Perry and Hayes [2] give an extensive list of factors assembled from several sources, and classified in terms of risks retainable by contractors, consultants and clients. Abdou [3] classified construction risks into three groups, i.e. construction finance, construction time and construction design. Shen [4] identified eight major risks accounting for project delay and ranked them based on a questionnaire survey with industry practitioners. Tah and Carr [5] classified project risks by using the hierarchical risk breakdown structure (HRBS) and classified them into internal and external risks. Chapman [6] grouped risks into four subsets: environment, industry, client and project. Shen [7] categorized them into six groups in accordance with the nature of the risks, i.e. financial, legal, management, market, policy and political. Chen et.al. [8] proposed 15 risks concern with project cost and divided them into three groups: resource factors, management factors and parent factors. Assaf and Al-Hejji [9] mentioned the risk factors as the delay factors in construction projects. Dikmen et al [10] used influence diagrams to define the factors which have influence on project risks. Zeng et al. [11] classified risk factors as human, site, material and equipment factors. Based on the above literature review, we propose risk classification as shown in figure 1.

3.0   RISK FACTOR PRIORITIZATION AND SELECTION

After classifying the inherent risks in construction projects, it is very important to select and prioritize the risk items in order to have an efficient risk management plan. Since we have a finite number of criteria and infinite number of feasible alternatives, the multiple criteria decision making model should be utilized. The main factors that taken into consideration in mentioned model are decision makers influence weights, their preferences for risk factor selection and the criteria for assessing risks. Group members consist of different experts in construction industry with variety in experience, knowledge and expertise. Experts with higher degree of competence should be assigned higher weights. Experts may not know or consider all the relevant information for a decision problem. To conquer this subject, an uncertainty factor named preference of every decision maker and related belief matrices are considered.

To apply this model, risk factor classification, projects requirements and objectives should be determined. Experts select the risk factors and then rank them to select N of them. Risk assessment and ranking criteria will be nominated by group members and finally T criteria will be used. To incorporate human inconsistency in decisions, it is suggested that all group members corporate in group aggregation process to ensure that the disparate individuals come to share the same decision objectives. Any individual role in a decision process, a preference for alternatives, and a judgment for assessment criteria are often expressed by linguistic terms as normal, more important. To deal with these uncertain and vague terms, crisp mathematical approaches cannot be applied. To handle these uncertainties, inaccurate and vague linguistic terms, the fuzzy logic is applied. The theory of fuzzy sets provides a framework and offers a calculus to address these fuzzy statements.

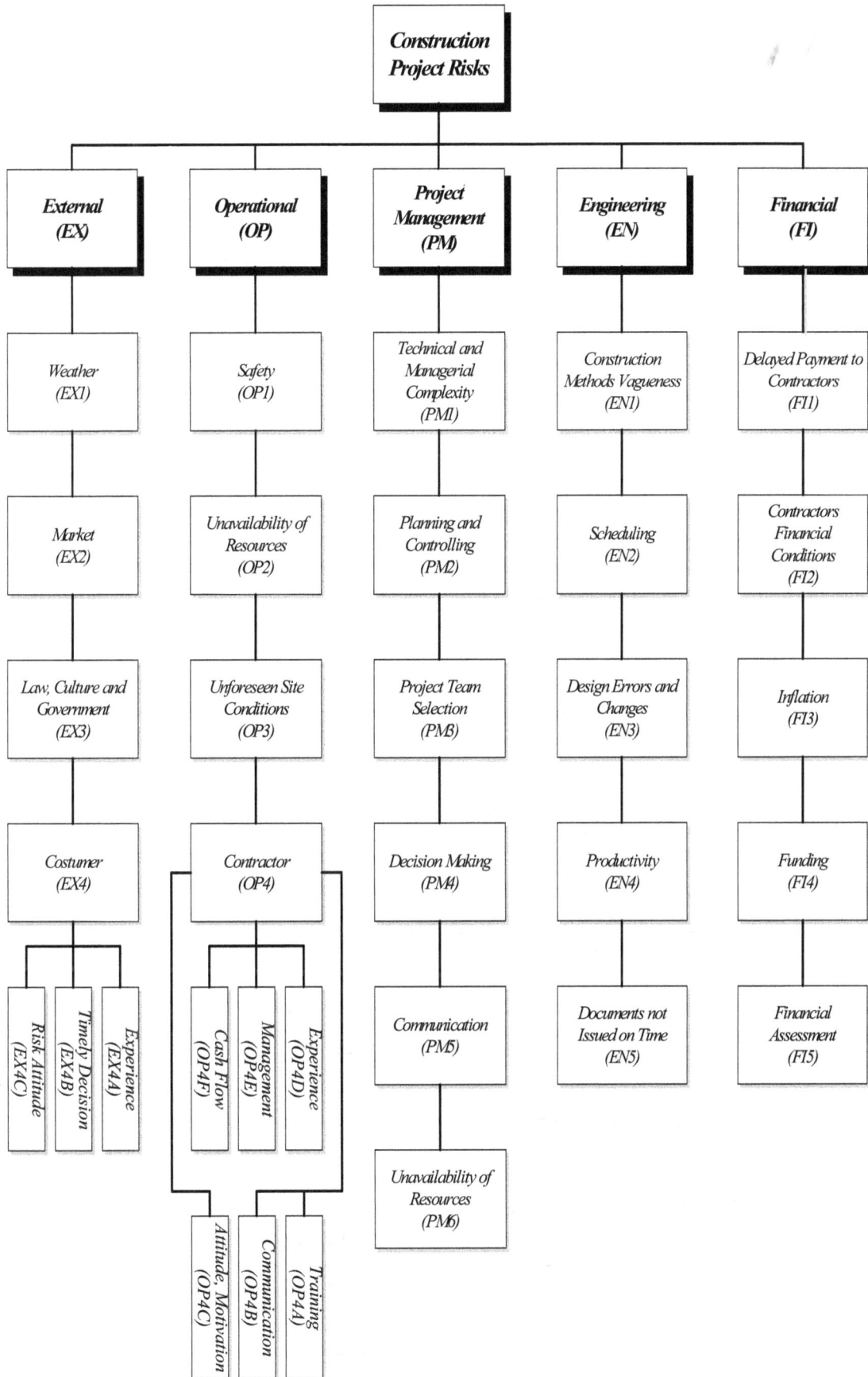

Figure 1: Construction Risk Classification

3.1   METHODOLOGY

Let $P = \{P_1, P_2, ..., P_n\}, n \geq 2$ be a given number of experts in the decision making group to prioritize and select risks from classified risk factors. The proposed model has ten steps within three stages:

Stage 1: Risk factor, assessment criteria and experts` influence weights determination

Step 1: By proposing classified risks in a group, every expert may have one or several possible risk factor selection. Through discussions and summarizations, $S = \{S_1, S_2, ..., S_m\}, m \geq 2$ is selected from alternative pool as final risk factors (alternatives) for prioritization.

Step 2: A criterion pool is constructed in this step and every members` assessment criteria is put into this pool. Each expert can propose his own assessment criteria for ranking and assessing the risk factors in this pool. Top T criteria, $C = \{C_1, C_2, ..., C_t\}$ are chosen as assessment criteria for risk selection problem.

Step 3: To consider the experience, knowledge and expertise of each expert, an influence weight is described and assigned to every expert. These influence weights are described by linguistic term $\tilde{v}_k, k = 1, 2, ..., n$. These weights can be determined through discussions in group or assigned by the leader of decision making group. These weights are assigned before or at the beginning of decision process. Table 1 shows related linguistic terms of decision makers. These linguistic terms and related membership functions are shown in figure 2. Triangular fuzzy numbers are used to map the linguistic terms to their corresponding fuzzy numbers. Table 2 presents a suggestive construction expert board to deal with risk selection in construction projects.

Table 1: Linguistic terms for describing weights of decision makers

Linguistic Terms	Membership Functions	Fuzzy Numbers	Supporting Intervals	Abbreviation
Normal	5x	(0,0.2,0.4)	$0 \leq x \leq 0.2$	c1
Normal	2-5x	(0,0.2,0.4)	$0.2 \leq x \leq 0.4$	c1
Important	5x-1	(0.2,0.4,0.6)	$0.2 \leq x \leq 0.4$	c2
Important	3-5x	(0.2,0.4,0.6)	$0.4 \leq x \leq 0.6$	c2
More Important	5x-2	(0.4,0.6,0.8)	$0.4 \leq x \leq 0.6$	c3
More Important	4-5x	(0.4,0.6,0.8)	$0.6 \leq x \leq 0.8$	c3
Most Important	5x-3	(0.6,0.8,1)	$0.6 \leq x \leq 0.8$	c4
Most Important	5-5x	(0.6,0.8,1)	$0.8 \leq x \leq 1$	c4

Table 2: Suggestive construction expert board in decision group

Experts	Linguistic Terms	Abbreviation
Construction Manager	Most Important	c4
Senior Execution Engineer	More Important	c3
Senior Design Engineer	More Important	c3
Site Engineer with 15 Years Experience	Important	c2
Expert Presented By Client	Normal	c1

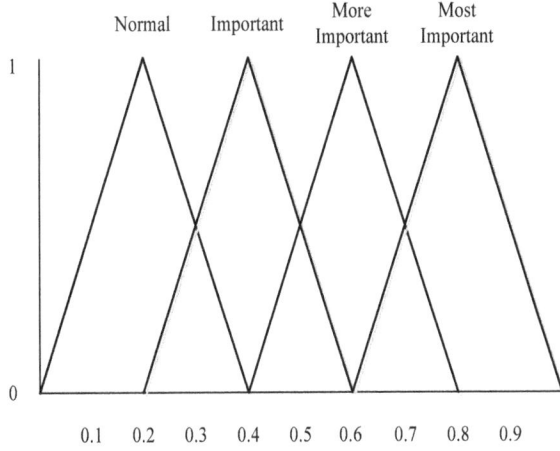

Figure 2: M.F. of decision makers weights

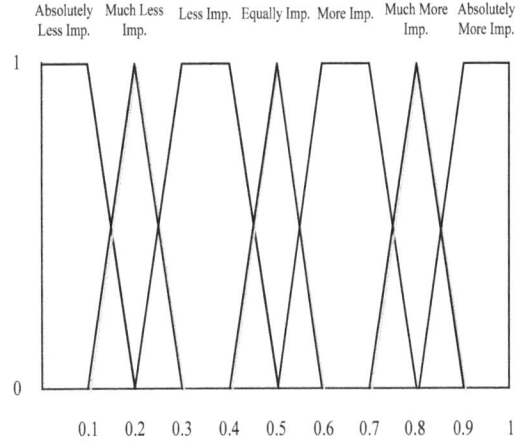

Figure 3: M.F. of assessment criteria comparison

Stage 2: Expert preference generation

Step 4: In this step each expert by using a pair-wise comparison expresses his opinion about outcomes of step 2. At first, a pair-wise comparison matrix $E = \left[\tilde{e}_{ij}^{k} \right]_{t \times t}$ is established. Every member of this matrix represents the quantified judgments on pairs of assessment criteria C_i and C_j $(i, j = 1, 2, ..., t, i \neq j)$. The linguistic terms and corresponding membership values which will be used for the comparison of the assessment criteria are described in Table 3 and figure 3. By utilizing the political model in this hybrid system, there is no obligation for experts to compare all the outcomes. Where ever the experts do not know or cannot compare the relative importance of assessment criteria C_i and C_j a '*' sign will be placed in pair-wise comparison matrix. By using following linguistic inference rules, the inconsistency of each pair-wise comparison matrix $E = \left[\tilde{e}_{ij}^{k} \right]_{t \times t}$ is corrected:

Rule 1: Positive-Transitive rule;

$\quad$ If $\tilde{e}_{ij}^{k} = a_s$ $(s = 4, 5, 6, 7)$ and $\tilde{e}_{jm}^{k} = a_t$ $(t = 4, 5, 6, 7)$, then $\tilde{e}_{im}^{k} = a_{\max(s,t)}$.

Rule 2: Negative-Transitive rule;

$\quad$ If $\tilde{e}_{ij}^{k} = a_s$ $(s = 3, 2, 1)$ and $\tilde{e}_{jm}^{k} = a_t$ $(t = 3, 2, 1)$, then $\tilde{e}_{im}^{k} = a_{\min(s,t)}$.

Rule 3: De-In-Uncertainty rule;

$\quad$ If $\tilde{e}_{ij}^{k} = a_s$ $(s = 4, 5, 6, 7)$ and $\tilde{e}_{jm}^{k} = a_t$ $(t = 3, 2, 1)$ or '*', then $\tilde{e}_{im}^{k} = a_i$ for any $t \leq i \leq s$ or '*'.

Rule 4: In-De-Uncertainty rule;

$\quad$ If $\tilde{e}_{ij}^{k} = a_s$ $(s = 3, 2, 1)$ or '*', and $\tilde{e}_{jm}^{k} = a_t$ $(t = 4, 5, 6, 7)$, then $\tilde{e}_{im}^{k} = a_i$ for any $s \leq i \leq t$ or '*'.

After calculating the comparison matrix $E = \left[\tilde{e}_{ij}^{k} \right]_{t \times t}$ by using the geometric mean of each row, consistent weights w_i^k $(i = 1, 2, ..., t)$ for every risk selection criterion is calculated. Resulting fuzzy numbers are normalized and described as

$$\tilde{w}_i^k = \frac{w_i^k}{\sum_{i=1}^{t} w_{i_0}^{k^R}} \, for \, i = 1, 2, ..., t; k = 1, 2, ..., n, \quad \tilde{w}_i^k \in F_T^*(R).$$

Table 3: Linguistic terms for the comparison of assessment criteria

Linguistic Terms	Membership Functions	Fuzzy Numbers	Supporting Intervals	Abbreviation
Absolutely Less Important	0	(0,0,0.1,0.2)	x=0	a1
	1		$0 \leq x \leq 0.1$	
	2-10x		$0.1 \leq x \leq 0.2$	
Much Less Important	10x-1	(0.1,0.2,0.2,0.3)	$0.1 \leq x \leq 0.2$	a2
	3-10x		$0.2 \leq x \leq 0.3$	
Less Important	10x-2	(0.2,0.3,0.4,0.5)	$0.2 \leq x \leq 0.3$	a3
	1		$0.3 \leq x \leq 0.4$	
	5-10x		$0.4 \leq x \leq 0.5$	
Equally Important	10x-4	(0.4,0.5,0.5,0.6)	$0.4 \leq x \leq 0.5$	a4
	6-10x		$0.5 \leq x \leq 0.6$	
More Important	10x-5	(0.5,0.6,0.7,0.8)	$0.5 \leq x \leq 0.6$	a5
	1		$0.6 \leq x \leq 0.7$	
	8-10x		$0.7 \leq x \leq 0.8$	
Much More Important	10x-7	(0.7,0.8,0.8,0.9)	$0.7 \leq x \leq 0.8$	a6
	9-10x		$0.8 \leq x \leq 0.9$	
Absolutely More Important	10x-8	(0.8,0.9,1,1)	$0.8 \leq x \leq 0.9$	a7
	1		$0.9 \leq x \leq 1$	
	0		x=1	

Step 5: To express the possibility of selecting a risk factor by experts, a belief level is introduced. The belief level b_{ij}^k ($i = 1,2,...,t, j = 1,2,...,m, k = 1,2,...,n$) belongs to a set of linguistic terms that contain various degrees of preferences required by decision makers. Where ever an expert do not know or cannot give a belief level a '**' sign is used in belief matrix. The linguistic terms for preference belief levels of alternatives are described in table 4.

Table 4: Linguistic terms for preference belief levels for alternatives

Linguistic Terms	Membership Functions	Fuzzy Numbers	Supporting Intervals	Abbreviation
Lowest	0	(0,0,0.1,0.2)	x=0	b1
	1		$0 < x < 0.1$	
	2-10x		$0.1 \leq x \leq 0.2$	
Very Low	10x-1	(0.1,0.2,0.2,0.3)	$0.1 \leq x \leq 0.2$	b2
	3-10x		$0.2 \leq x \leq 0.3$	
Low	10x-2	(0.2,0.3,0.4,0.5)	$0.2 \leq x \leq 0.3$	b3
	1		$0.3 \leq x \leq 0.4$	
	5-10x		$0.4 \leq x \leq 0.5$	
Medium	10x-4	(0.4,0.5,0.5,0.6)	$0.4 \leq x \leq 0.5$	b4
	6-10x		$0.5 \leq x \leq 0.6$	
High	10x-5	(0.5,0.6,0.7,0.8)	$0.5 \leq x \leq 0.6$	b5
	1		$0.6 \leq x \leq 0.7$	
	8-10x		$0.7 \leq x \leq 0.8$	
Very High	10x-7	(0.7,0.8,0.8,0.9)	$0.7 \leq x \leq 0.8$	b6
	9-10x		$0.8 \leq x \leq 0.9$	
Highest	10x-8	(0.8,0.9,1,1)	$0.8 \leq x \leq 0.9$	b7
	1		$0.9 \leq x \leq 1$	
	0		x=1	

Step 6: By applying the normalized weights resulted from step 4 into belief level matrix $(b_{ij}^k)(k = 1,2,...,n)$ and aggregate the results, belief vectors $\overline{b}_j^k = \tilde{w}_{j_1}^k * b_{jj_1}^k + \tilde{w}_{j_2}^k * b_{jj_2}^k + ... + \tilde{w}_{j_s}^k * b_{jj_s}^k$ where $b_{jj_i}^k$ $(i = 1,2,...,s)$ *is not* '**' are obtained.

Step 7: At this step, normalized weight of decision maker is calculated.

$$\tilde{v}_k^* = \frac{\tilde{v}_k}{\sum_{i=1}^n v_{i_0}^R} \, for \, k = 1,2,...,n.$$

Step 8: By applying the normalized weight obtained from previous step and belief vectors obtained from step 6, a weighted normalized fuzzy decision matrix is constructed.

$$(\tilde{r}_1, \tilde{r}_2,...,\tilde{r}_m) = (\tilde{v}_1^*, \tilde{v}_2^*,...,\tilde{v}_n^*) \begin{pmatrix} \overline{b}_1^1 & \overline{b}_2^1 & \cdots & \overline{b}_m^1 \\ \overline{b}_1^2 & \overline{b}_2^2 & \cdots & \overline{b}_m^1 \\ \vdots & \vdots & \ddots & \vdots \\ \overline{b}_1^n & \overline{b}_2^n & \cdots & \overline{b}_m^n \end{pmatrix} \quad where \; \tilde{r}_j = \sum_{k=1}^n \tilde{v}_k^* \overline{b}_j^k .$$

Step 9: The ideal solution is assessed and the distance between alternatives (risk factor) and the ideal solution will be calculated. Alternative (risk factor) with the least distance is assumed to be the highest priority risk factor selected by group decision.

Suppose elements in decision matrix defined as $\tilde{r}_m = (r_m^L, r_m^M, r_m^R)$ and the ideal alternative is named $A^* = [\tilde{x}_j^*] : \tilde{b}_j^* = (x_j^{*L}, x_j^{*M}, x_j^{*R})$. The distance between every alternative in decision matrix and ideal alternative is calculated as follow:

$$d_i = d_{(\tilde{r}_m, A^*)} = \sqrt{\frac{1}{3}\sum_{j=1}^m \left[\left(r_m^L - x_j^{*L} \right)^2 + \left(r_m^M - x_j^{*M} \right)^2 + \left(r_m^R - x_j^{*R} \right)^2 \right]}$$

Assume that decision matrix is a set of pairs (r_K, r_L) that r_K is preferred to r_L. This implies that risk factor K has more effect on project objectives than risk factor L and distance (d_i) between risk factor K to ideal set of alternatives (risk items) is less than risk factor L $(d_L \geq d_K)$. As we stated before, experts may have no or incomplete information about assessment criteria; so we the human errors in prediction should be considered. This error (d^-) and the amount of incredibility (error) in pair-wise comparison of alternatives (B) to find the negative ideal solution is defined as bellow:

$$d_{K,L}^- = \begin{cases} d_K - d_L & d_K > d_L \\ 0 & d_K \leq d_L \end{cases}$$

$$d_{K,L}^- = \max\{0, d_K - d_L\}$$

$$B = \sum_{(K,L) \in \tilde{r}_m} d_{K,L}^-$$

To obtain the positive ideal solution, a new value called credibility judgment degree is defined between two risk factors K and L.

$$d_{K,L}^+ = \begin{cases} d_L - d_K & d_L > d_K \\ 0 & d_L \leq d_K \end{cases}$$

$$d_{K,L}^+ = \max\{0, d_L - d_K\}$$

$$G = \sum_{(K,L) \in \tilde{r}_m} d_{K,L}^+$$

To obtain the final ideal solution, credibility degree should be maximized while incredibility (error) degree should be minimized. Amount of this difference (h) and P should be defined by decision makers $(G - B \geq h)$. The membership function of this ideal solution is as follow:

$$\mu_{(G-B)} = \frac{(G-B)-(h-P)}{P} = \frac{\sum_{(K,L \in \tilde{r}_m)} (d_L - d_K)-(h-P)}{P}$$

In the field of risk selection in construction projects, h can be the defined as the least effect of a risk item in project objective and amount of P can be described as the highest effect of a risk item. The membership function of $G-B$ is shown on figure 4.

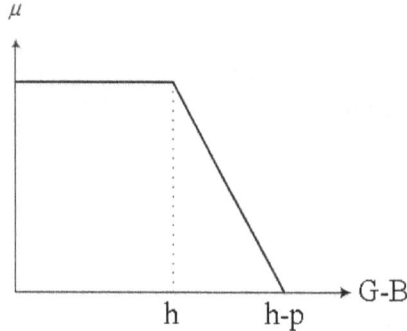

Figure 4: Membership Function of G-B

The distance (d_i) of alternatives (risk factors) with ideal solution (G-B) is calculated. The risk factor with the least distance is selected as the highest priority factor to be considered and other factors will be ranked in ascending order.

4.0 COMPARING THE PROPOSED FUZZY MCDM MODEL WITH FUZZY AHP

In this section, a comparison between proposed fuzzy MCDM model and different fuzzy AHP approaches is presented. This part of the paper is followed by definition of AHP, Fuzzy AHP, their shortcomings and benefits of our model comparing to fuzzy AHP.

4.1 AHP

The AHP is a popular decision making technique that has proven easy to understand and plausible for prioritizing alternatives among multi-criteria and multi-attributes (Saaty, 1990, Kim, Whang, 1993, Cheng, 1996, Badri, 1999, Lee, Kwak, 1999, Harbi, 2001). The use of AHP need not involve troublesome mathematics but decomposition, pair-wise comparison and priority vector creation (Zeng et.al. 1997). Because AHP does not take into account the uncertainty associated with the mapping of one's judgment to a number and also the subjective judgments, selection, and preference of decision makers exert a strong influence in the AHP. AHP method can only deal with definite scales in reality (Zeng et.al. 1997) while Construction problems are complicated usually involving massive uncertainties and subjectivities. In a typical AHP method, experts have to give a definite number within a 1–9 scale to the pair-wise comparison so that the priority vector can be computed. However factor comparisons often involve certain amount of uncertainty and subjectivity because sometimes, experts cannot compare two factors due to the lack of adequate information. In this case, a typical AHP method has to be discarded due to the existence of fuzzy or incomplete comparisons. In this case a fuzzy AHP approach may be applied.

4.2 FUZZY AHP

A Fuzzy AHP is an important extension of the typical AHP method which was first introduced by Laarhoven and Pedrycz. One of the drawbacks of fuzzy AHP method is the complicated fuzzy operation and the lack of proven techniques to address fuzzy consistency and fuzzy priority vector.

4.3 COMPARISON OF PROPOSED FUZZY MCDM MODEL WITH FUZZY AHP

To discover the characteristics and advantages of proposed fuzzy MCDM model and fuzzy AHP a comparison between Main characteristics, advantages and disadvantages of different fuzzy AHP approaches (Tuysuz, Kahraman 2006) is implemented in Table 5.

Table 5: The comparison of different fuzzy AHP methods with proposed fuzzy MCDM

Source	The main characteristics of method	Advantages (+) and Disadvantages (-)
Laarhoven,, Pedrycz (1983)	• Direct extension of Saaty's AHP method with triangular fuzzy numbers • Lootsma's logarithmic least square method is used to derive fuzzy weights and fuzzy performance scores	(+) The opinions of multiple decision makers can be modeled in the reciprocal matrix. (-) There is not always a solution to the linear equations. (-)The computational requirement is tremendous, even for a small problem. (-) It allows only triangular fuzzy numbers to be used.
Buckley (1985)	• Extension of Saaty's AHP method with trapezoidal fuzzy numbers • Uses the geometric mean method to derive fuzzy weights and performance scores	(+) It is easy to extend to the fuzzy case. (+) It guarantees a unique solution to the reciprocal comparison matrix. (-) The computational requirement is tremendous.
Boender, Grann, Lootsma (1989)	• Modifies van Laarhoven and Pedrycz's method • Presents a more robust approach to the normalization of the local priorities	(+) The opinions of multiple decision makers can be modeled. (-)The computational requirement is tremendous.
Chang (1996)	• Synthetical degree values low. • Layer simple sequencing • Composite total sequencing	(+) The computational requirement is relatively low. (+) It follows the steps of crisp AHP. It does not involve additional operations. (-) It allows only triangular fuzzy numbers to be used.
Cheng (1996)	• Builds fuzzy standards • Represents performance scores by membership functions both probability and possibility measures. • Uses entropy concepts to	(+) The computational requirement is not tremendous. (-) Entropy is used when probability distribution is known. The method is based on both probability and

	calculate aggregate weights	possibility measures.
Proposed Fuzzy MCDM	• Extension of rational model • Consensus rule based • Self optimization • Characterized for risk analysis • Uses Euclidean distance to find optimal solution • Pair-wise inconsistency correction	(+) Uncertainty factors in group decision making are assessed by applying fuzzy logic (+) Final solution is prioritized (+) Different fuzzy numbers and membership functions can be applied (+) Experts can have inconsistent evaluation (+) Experts decision weight is efficiently applied to model (-) The computation requirement is relatively high

5.0   CASE STUDY

To illustrate the application of proposed fuzzy multi-criteria group decision making model in construction risk selection, we applied this model to a typical construction project as a case study.

Suppose a group of experts to identify inherent risk in a construction project consist of three experts P1, P2 and P3. To avoid complexity of manual computations, it is assumed that experts have same influence weights. Their weights, preference for risk factor selection and judgments for proposed assessment criteria are described in table 1, 3 and 4. The risk selection process by using proposed method is described as follow:

Stage 1: Alternatives, assessment criteria and influence weights generation

Step 1: to initiate the selection process, involved risks in project should be classified. Each expert proposes one or more risk factor for project risk selection. Final alternative risk S is determined by merging similar risk factors.

$S = \{S_1, S_2, S_3, S_4\}$

S1: Safety, S2: Scheduling, S3: Unavailability of resources, S4: Weather

Step 2: The experts should assess these risk factors with regard to magnitude and effect on project objectives by proposing an assessment criteria. In this case study we put emphasis on project duration and assess risk factors based on their impact on project duration. By merging overlapped criteria, five assessment criteria C1, C2, C3, C4 and C5 are obtained.

C1: Effect of new safety plans on project duration

C2: The impact of changing operations` scheduling on project delivery

C3: Change operations from non-critical to critical due to unavailability of resources

C4: Consequence of undesired weather condition on project delays with regard to project location.

C5: Impact of risk factor on costumer

Step 3: to avoid the complexity, we assume that all experts have same influence weights as 'normal'.

Stage 2: Individual preferences generation

Step 4: Five assessment criteria obtained from previous step are being judged by using pair-wise comparison. At this step, every expert should present his individual judgment for assessment criteria. Resulted pair-wise comparison matrices are calculated as follow:

$$E^1 = E^2 = E^3 = \begin{pmatrix} EI & EI & * & * & EI \\ EI & EI & * & EI & * \\ * & * & EI & * & * \\ * & EI & * & EI & EI \\ EI & * & * & EI & EI \end{pmatrix} = \begin{pmatrix} a_4 & a_4 & * & * & a_4 \\ a_4 & a_4 & * & a_4 & * \\ * & * & a_4 & * & * \\ * & a_4 & * & a_4 & a_4 \\ a_4 & * & * & a_4 & a_4 \end{pmatrix}$$

To correct the inconsistency of each pair-wise comparison matrix, the positive-transitive, De-In and In-De uncertainty rules are applied. Finalized pair-wise comparison matrices to express the possibility of selecting a risk factor, under certain criteria is as follow:

$$E^1 = E^2 = E^3 = \begin{pmatrix} EI & EI & * & EI & EI \\ EI & EI & * & EI & * \\ * & * & EI & * & * \\ EI & EI & * & EI & EI \\ EI & * & * & EI & EI \end{pmatrix} = \begin{pmatrix} a_4 & a_4 & * & a_4 & a_4 \\ a_4 & a_4 & * & a_4 & * \\ * & * & a_4 & * & * \\ a_4 & a_4 & * & a_4 & a_4 \\ a_4 & * & * & a_4 & a_4 \end{pmatrix}$$

Normalized pair-wise comparison matrix and consistent weight for every assessment criteria are calculated by computing the geometric mean of every row.

$$\begin{pmatrix} w_1^1 \\ w_2^1 \\ w_3^1 \\ w_4^1 \\ w_5^1 \end{pmatrix} = \begin{pmatrix} w_1^2 \\ w_2^2 \\ w_3^2 \\ w_4^2 \\ w_5^2 \end{pmatrix} = \begin{pmatrix} w_1^3 \\ w_2^3 \\ w_3^3 \\ w_4^3 \\ w_5^3 \end{pmatrix} = \begin{pmatrix} \sqrt[4]{a_4^4} \\ \sqrt[3]{a_4^3} \\ a_4 \\ \sqrt[4]{a_4^4} \\ \sqrt[3]{a_4^3} \end{pmatrix} = \begin{pmatrix} a_4 \\ a_4 \\ a_4 \\ a_4 \\ a_4 \end{pmatrix} = \begin{pmatrix} [10x-4,6-10x] \\ [10x-4,6-10x] \\ [10x-4,6-10x] \\ [10x-4,6-10x] \\ [10x-4,6-10x] \end{pmatrix}$$

$$\sum_{i=1}^{5} w_{i_0}^{1^R} = \sum_{i=1}^{5} w_{i_0}^{2^R} = \sum_{i=1}^{5} w_{i_0}^{3^R} = 3$$

$$\begin{pmatrix} \tilde{w}_1^1 \\ \tilde{w}_2^1 \\ \tilde{w}_3^1 \\ \tilde{w}_4^1 \\ \tilde{w}_5^1 \end{pmatrix} = \begin{pmatrix} \tilde{w}_1^2 \\ \tilde{w}_2^2 \\ \tilde{w}_3^2 \\ \tilde{w}_4^2 \\ \tilde{w}_5^2 \end{pmatrix} = \begin{pmatrix} \tilde{w}_1^3 \\ \tilde{w}_2^3 \\ \tilde{w}_3^3 \\ \tilde{w}_4^3 \\ \tilde{w}_5^3 \end{pmatrix} = \frac{1}{3} \begin{pmatrix} a_4 \\ a_4 \\ a_4 \\ a_4 \\ a_4 \end{pmatrix}$$

Step 5: To express the possibility of selecting a risk factor (S_i) under criterion (C_j), three belief level matrices are obtained by group members:

$$\begin{pmatrix} b_{11}^1 & b_{12}^1 & b_{13}^1 & b_{14}^1 & b_{15}^1 \\ b_{21}^1 & b_{22}^1 & b_{23}^1 & b_{24}^1 & b_{25}^1 \\ b_{31}^1 & b_{32}^1 & b_{33}^1 & b_{34}^1 & b_{35}^1 \\ b_{41}^1 & b_{42}^1 & b_{43}^1 & b_{44}^1 & b_{45}^1 \end{pmatrix} = \begin{pmatrix} M & VL & ** & ** & ** \\ VH & M & ** & ** & ** \\ ** & ** & M & VL & ** \\ ** & VL & ** & ** & M \end{pmatrix} = \begin{pmatrix} b_4 & b_1 & ** & ** & ** \\ b_7 & b_4 & ** & ** & ** \\ ** & ** & b_4 & b_1 & ** \\ ** & b_1 & ** & ** & b_4 \end{pmatrix},$$

$$\begin{pmatrix} b_{11}^2 & b_{12}^2 & b_{13}^2 & b_{14}^2 & b_{15}^2 \\ b_{21}^2 & b_{22}^2 & b_{23}^2 & b_{24}^2 & b_{25}^2 \\ b_{31}^2 & b_{32}^2 & b_{33}^2 & b_{34}^2 & b_{35}^2 \\ b_{41}^2 & b_{42}^2 & b_{43}^2 & b_{44}^2 & b_{45}^2 \end{pmatrix} = \begin{pmatrix} M & VL & ** & ** & ** \\ VH & ** & M & ** & ** \\ ** & ** & VL & M & ** \\ ** & M & ** & ** & VL \end{pmatrix} = \begin{pmatrix} b_4 & b_1 & ** & ** & ** \\ b_7 & ** & b_4 & ** & ** \\ ** & ** & b_1 & b_4 & ** \\ ** & b_4 & ** & ** & b_1 \end{pmatrix},$$

$$\begin{pmatrix} b_{11}^3 & b_{12}^3 & b_{13}^3 & b_{14}^3 & b_{15}^3 \\ b_{21}^3 & b_{22}^3 & b_{23}^3 & b_{24}^3 & b_{25}^3 \\ b_{31}^3 & b_{32}^3 & b_{33}^3 & b_{34}^3 & b_{35}^3 \\ b_{41}^3 & b_{42}^3 & b_{43}^3 & b_{44}^3 & b_{45}^3 \end{pmatrix} = \begin{pmatrix} VL & M & ** & ** & ** \\ M & ** & VH & ** & ** \\ ** & ** & M & VL & ** \\ ** & M & ** & ** & VL \end{pmatrix} = \begin{pmatrix} b_1 & b_4 & ** & ** & ** \\ b_4 & ** & b_7 & ** & ** \\ ** & ** & b_4 & b_1 & ** \\ ** & b_4 & ** & ** & b_1 \end{pmatrix}.$$

Step 6: By applying the results obtained from step 4 to belief level matrix, three belief vectors are obtained as follow:

$$\bar{b}_1^1 = \frac{1}{3}\left(a_4^2 + a_4 a_1\right), \bar{b}_2^1 = \frac{1}{3}\left(a_4^2 + a_4 a_7\right), \bar{b}_3^1 = \frac{1}{3}\left(a_4^2 + a_4 a_1\right), \bar{b}_4^1 = \frac{1}{3}\left(a_4^2 + a_4 a_1\right),$$

$$\bar{b}_1^2 = \frac{1}{3}\left(a_4^2 + a_4 a_1\right), \bar{b}_2^2 = \frac{1}{3}\left(a_4^2 + a_4 a_1\right), \bar{b}_3^2 = \frac{1}{3}\left(a_4^2 + a_4 a_1\right), \bar{b}_4^2 = \frac{1}{3}\left(a_4^2 + a_4 a_1\right),$$

$$\bar{b}_1^3 = \frac{1}{3}\left(a_4^2 + a_4 a_1\right), \bar{b}_2^3 = \frac{1}{3}\left(a_4^2 + a_4 a_7\right), \bar{b}_3^3 = \frac{1}{3}\left(a_4^2 + a_4 a_1\right), \bar{b}_4^3 = \frac{1}{3}\left(a_4^2 + a_4 a_1\right).$$

Stage 3: Group aggregation
Step 7: The normalized weight of decision makers denoted as follow:

$$v_1 = v_2 = v_3 = c_1$$

$$\sum_{i=1}^{3} v_{i_0}^R = 1.2$$

$$v_1^* = v_2^* = v_3^* = \frac{1}{1.2} a_4$$

Step 8: By applying obtained results from steps 6 and 7, weighted and normalized fuzzy decision vector is constructed:

$$\tilde{r}_1 = v_1^* \bar{b}_1^1 + v_2^* \bar{b}_1^2 + v_3^* \bar{b}_1^3 = \frac{1}{1.2} a_4^2 \left(a_4 + a_1\right) = \frac{1}{1.2}\left[(10x-4)^2, (6-10x)^2\right]\left[(10x-4),(8-20x)\right],$$

$$\tilde{r}_2 = v_1^* \bar{b}_2^1 + v_2^* \bar{b}_2^2 + v_3^* \bar{b}_2^3 = \frac{1}{1.2} a_4^2 \left(a_4 + a_7\right) = \frac{1}{1.2}\left[(10x-4)^2, (6-10x)^2\right]\left[(20x-12),(6-10x)\right],$$

$$\tilde{r}_3 = v_1^* \bar{b}_3^1 + v_2^* \bar{b}_3^2 + v_3^* \bar{b}_3^3 = \frac{1}{1.2} a_4^2 \left(a_4 + a_1\right) = \frac{1}{1.2}\left[(10x-4)^2, (6-10x)^2\right]\left[(10x-4),(8-20x)\right],$$

$$\tilde{r}_4 = v_1^* \bar{b}_4^1 + v_2^* \bar{b}_4^2 + v_3^* \bar{b}_4^3 = \frac{1}{1.2} a_4^2 \left(a_4 + a_1\right) = \frac{1}{1.2}\left[(10x-4)^2, (6-10x)^2\right]\left[(10x-4),(8-20x)\right].$$

Step 9: To reach the ideal solution, it is assumed that the ideal risk factor has minimum 0.25 and maximum 0.75 effect on project duration. The distances between obtained decision vector item for each risk factor and ideal risk factor are depicted below:

$$d_{S_1} = 0.1536 \quad (Safety)$$
$$d_{S_2} = 0.0695 \quad (Scheduling)$$
$$d_{S_3} = 0.0725 \quad (Unavailability\ of\ resources)$$
$$d_{S_4} = 0.1536 \quad (Weather)$$

5.1   DISCUSSION OF RESULTS

By considering relative Euclidean distance, it is concluded that 'scheduling' risk factor has the most effect on project duration and 'unavailability of resources', 'safety' and 'weather' are on next order. Another conclusion that can be obtained from these results is the criticality and dependency of "Scheduling" and "Unavailability of resources". As can be seen, "Unavailability of resources" has a closer distance to the most critical risk factor than "Safety" and "Weather" which shows a dependency between "Unavailability of resources" and "Scheduling". Due to the dependency of these two risk factors, improving them should be done simultaneously. Otherwise improving one risk factor may lead to criticality of other.

Considering the result of this case study, project manager or decision maker should consider factors and operations that may cause "scheduling" to be critical on project objective.

For instance, he may re-arrange the float times or make revisions on critical paths. Also he may take into consideration the share activities that overlap the "Unavailability of resources".

5.2   RESULT COMPARISON WITH FUZZY AHP

To discuss the difference between the proposed fuzzy MCDM and the fuzzy AHP, same case study has been implemented using Chang (1996) fuzzy AHP approach. Because of the advantages Chang's extent analysis on fuzzy AHP are relatively superior to the others due to the reasons mentioned in Table 5, this method will be used in project risk evaluation (Tuysuz, Kahraman 2006). Because Chang`s approach allows only triangular fuzzy numbers, related non-triangular fuzzy numbers in case study, has been converted to triangular fuzzy numbers. After relatively high and time consuming computations, obtained results are as follow:

Risk Factor 1 = *Scheduling*
Risk Factor 2 = *Unavailability of resources*
Risk Factor 3 = *Safety*
Risk Factor 4 = *Weather*

5.3   DISCUSSION

As concluded from this comparison, the priority rank of risk factors is same with proposed fuzzy MCDM method but the computations in utilized fuzzy AHP method is relatively high and limitation in applying other membership functions and fuzzy numbers rather than triangular fuzzy numbers, make it impractical in the field of construction risk assessment. Also there is no rational comparison between prioritized risk factors and as the result risk mitigation strategy cannot effectively be added to risk management process.

6.0   CONCLUSIONS

In this paper we introduced a comprehensive hierarchical risk classification for construction projects through an extensive literature review and experiences in different projects. The main matter in an effective risk management plan is managing the most effective risks which have the maximum effect on project objectives. Due to lack of information and limited time, all the risk factors in a project cannot be considered for assessment. So a comprehensive risk selection mechanism should be developed to prioritize the inherent risks. In this study we developed this mechanism through a fuzzy multi criteria decision making model which is based on group decision making. Presented method has both advantages of a self optimization and no limitation for experts. Case studies have shown reasonable results by utilizing this method. As shown in case study results, not only prioritized risk factors can be selected by proposed method but also the interdependency of risk factors can be identified by comparing the relative distance of risk factors to each other. This option gives the decision makers a guide map of managing relative risk factors otherwise improving one factor will make others be critical. Several methods presented to solve above MCDM problems. Some of them are based on ideal alternative in the decision maker's opinion such as TOPSIS and ELECTRE. In the cases where ideal alternative and weight of criteria are not available for decision maker, aforesaid methods are not applicable. One of the shortcomings of this method is the tedious calculations of matrices. This can be improved by programming the calculations using spreadsheet or other programming solutions. Also in this study to simplify the fuzzy sets, we utilized the triangular fuzzy membership functions that may not be suitable for complex systems. Further studies can be conducted in developing the programming solution for this model and utilizing other membership functions for complex problems.

REFERENCES

[1] Project Management Institute, (2008) A guide to the project management body of knowledge. Project Management Institute Standards Committee.

[2] Perry, J.H. and Hayes, R.W. (1985) Risk and Its Management in Construction Projects, Proceedings of the Institution of Civil Engineering, Part I, 78, 499-521.

[3] Abdou, O.A. (1996) Managing Construction Risks, Journal of Architectural Engineering, 2(1), 3-10.

[4] Shen, L.Y., Wu, G.W.C. and Ng, C.S.K. (2001) Risk Assessment for Construction Joint Ventures in China, Journal of Construction Engineering and Management, 127(1), 76-81.

[5] Tah J.H.M and Carr V., A proposal for construction project risk assessment using fuzzy logic,Construction Management and Economics (2000) 18, 491-500.

[6] Chapman R.J. (2001) The Controlling Influences on Effective Risk Identification and Assessment for Construction Design Management, International Journal of Project Management, 19, 147-160.

[7] Shen, L.Y. (1997) Project Risk Management in Hong Kong, International Journal of Project Management, 15(2), 101-105.

[8] Chen, H., Hao, G., Poon, S.W. and Ng, F.F. (2004) Cost Risk Management in West Rail Project of Hong Kong, 2004 AACE International Transactions.

[9] Assaf, S. A. and Al-Hejji, S. (2006) Causes of delay in large construction projects, International Journal of Project Management, 24(4), 349-357.

[10] Dikmen, I., Birgonul, M., Han, S., (2007) Using fuzzy risk assessment to rate cost overrun risk in international construction projects, International Journal of Project Management, 25, 494–505.

[11] Zeng, J., An, M., and Smith, N. J. (2007) Application of a fuzzy based decision making methodology to construction project risk assessment, International Journal of Project Management, 25, 589–600.

10

ENERGY EFFICIENT GREEN BUILDING BASED ON GEO COOLING SYSTEM IN SUSTAINABLE CONSTRUCTION OF MALAYSIA

M.R. Alam[1], M.F.M.Zain[2] and A.B.M.A. Kaish[3]

[1, 2, 3] Faculty of Engineering and Built Environment, Universiti Kebangsaan Malaysia, Selangor, Malaysia

*Corresponding E-mail : arabiul07@yahoo.com

ABSTRACT

This paper focuses geo energy based cooling of building in tropical country like Malaysia where temperature rises in daytime and goes beyond to a comfortable limit. In this study inside temperature of the room in daytime is considered to be reduced by adding thermal conductivity media inside the room elements such as walls through its connectivity to the underground soil where temperature is less than the ambient room temperature. Due to the underground connectivity of thermal conductivity media a flow of heat creates from the room to the underground soil and tries to produce a thermal balance between these two medias and therefore, room temperature drops to a temperature close to the underground soil temperature. Aluminium pipes are considered as high thermal conductivity material. The entire study is done numerically using ANSYS 11 finite element software to determine the role of underground soil and thermal conductivity pipes in cooling of building rooms. In numerical investigation heat flow between two systems (building rooms equipped with thermal conductivity pipe and underground soil) is studied and the performance of the conductivity materials is examined. The room temperature in the presence of thermal conductivity media as well as mechanical cooling system is also investigated in this study. It is seen that high thermal conductivity media plays a role in transferring heat from room to the underground soil and makes cooling of the building effectively. It acts also effectively when it uses with other mechanical cooling system of the building.

Keywords: Geo energy, heat flow, thermal conductivity pipe, underground soil, mechanical cooling system

1.0 INTRODUCTION

Energy efficient building design is now a great concern among researchers and scientists as global energy demand is increasing gradually. Researchers have carried out researches regarding this issue in the last few years. Orientation, insulation of entire building envelope, WWR (window to wall ratio), double skin facades, high performance glazing, complete shading from direct sunlight, air-tightness and etc., were taken into consideration in their research interests for making building energy efficient.

Underground soil could act as a passive cooling media for building built in tropical environment due to its much lower temperature than the ambient temperature of the building and could play a role in saving cooling energy that requires for cooling of building in tropical or hot region. Available literature related to the underground soil temperature indicates that underground soil temperature varies from 8^0C to 27^0C in some parts of the world especially in cold dominated areas (North Americas) and 15^0C to 25^0C in tropical climate [1, 2, 3]. Nassar et al [4] carried out research on the evaluation of thermal properties of underground soil in Libya. They conducted their study on soil temperature at a depth of 4 m from the ground surface. They found average soil temperature of 21^0C at this depth of underground soil. Bansal et al [5] measured underground soil temperature at the same depth (4m) under the condition of wetted soil covered by dry black soil with the shaded surface over the year at New Delhi, India. They found maximum soil temperature of 17.5^{0}C at that depth of wetted soil. Outdoor temperature (shown in Figure 1) under mixed dipterocarp forests of peninsular Malaysia for a typical day of July to August 2006 determined by Nik et al [6]. They mentioned that temperature of soil under forest cover is consistently lower than of the open air by 4 to 6^{0}C due to shading effect of forest cover. Cui et al [7] found out temperature variation with depth of underground soil for a typical day of tropical countries (see Figure 2). Therefore, it is seen that underground soil temperature always will be less than the ambient temperature of the building if there is no effect of natural heat sources. Variation of underground soil temperature is usually found up to a depth of 3 m thereafter, it becomes constant. However, it depends on several factors such as: thermal conductivity, density and specific heat, climate of surrounding environment (meteorological conditions, solar radiation intensity, wind speed, rain, humidity and the air temperature), any natural heat sources near the earth crust, etc.

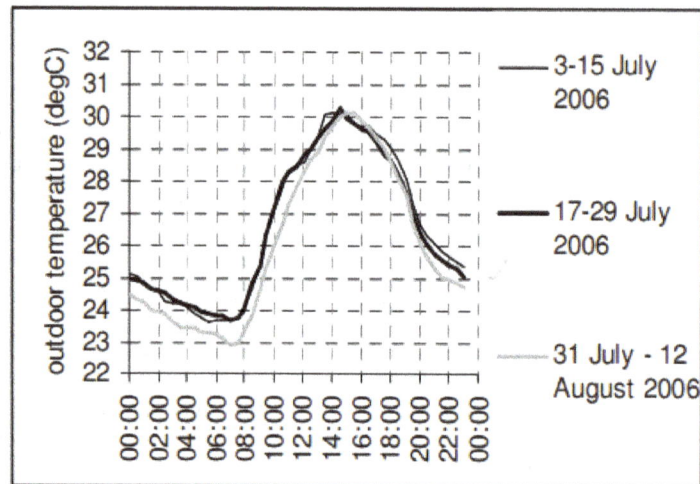

Figure 1: Outdoor temperature under mixed dipterocarp forests of peninsular Malaysia for a typical day of July to August 2006 [6]

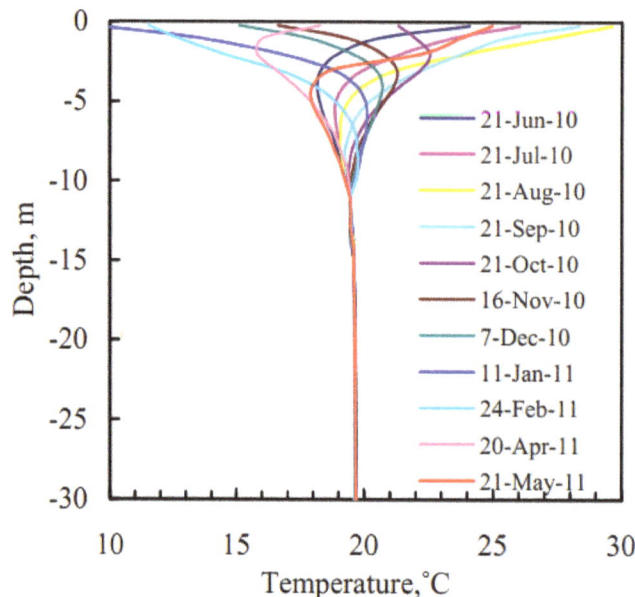

Figure 2: Temperature variation of underground soil with depth for typical days in Malaysia [7]

Using the advantages of underground soil temperature a case study of the cooltek house on ground cooling of ventilation air for energy efficient house in Malaysia was carried out by Reiman et. al. [8]. They analyzed an on-site measurements of the ground cooling air duct system of the energy efficient cooltek house in Melaka, Malysia. Indoor, outdoor and inside the ground cooling air duct temperature, humidity and CO_2 were measured in two modes of operations such as fully passive mode where ventilation is driven by the thermal pull of solar chimney and a hybrid mode where a small fan assists the solar chimney in ventilating the building. They mentioned that ground cooling system of inlet air helps to increase the energy efficiency of the house. Underground soil sometime also uses as heat sink for a modified air conditioner. Geothermal energy is also being used in other ways in heating and cooling of building such as Earth-air heat exchanger, Ground source heat pump, Bore hole heat exchanger, Energy piles, Ground absorbers, etc., in Germany, U.K, and other countries [9-11].

Temperature distribution inside the building is not constant during hot and cold time of day. It varies along vertically as well as horizontally inside of the room. Angela Simone and Carsten Rode [12] measure vertical and horizontal temperatures of the riso flexhouse room with different heating control principles. They found that the vertical temperature profile of the room shows notable temperature gradients, especially at the occupied zone when the solar gain is large, while otherwise the horizontal temperature distribution is smaller. They measured vertical temperatures at 1 m away from the window at heights of 0.1 m, 0.6 m, 1.1 m and 1.7 m above the floor from 8 am to 5 pm of a day. They found 3^0C to 5^0C vertical temperature variation in between 11.30 am to 2.00 pm. Yasui et al [13] carried our study on indoor thermal environment and vertical temperature gradient in large workshop with polycarbonate roofing of a school without air conditioning in summer time. They found large variation of vertical temperature inside the room. They mentioned that due to high solar radiation in summer and polycarbonate roofing system of the building indoor temperature becomes high and large vertical temperature gradient is formed. Pollard et al [14] found vertical temperature variation inside the room in between 0.5^0C and 1.5^0C for solar radiation in between 1 pm and 3 pm. Overby [15] also found similar variation of vertical temperature inside the room (shown in Figure 3).

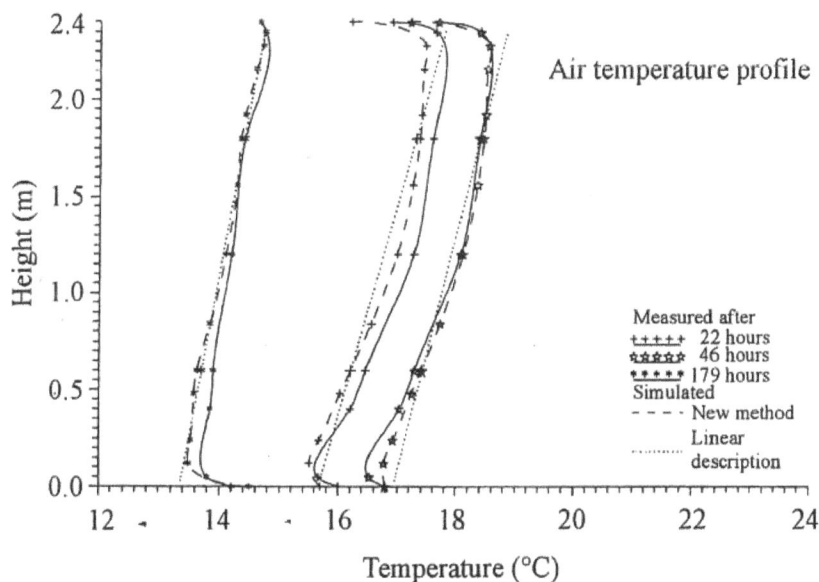

Figure 3: Vertical variation of temperature inside the building room [15]

In this study, a vertical variation of interior room temperature as well as variation of soil temperature up to a depth of 10 feet was taken into consideration during numerical analysis of heat transfer problem of the entire system. Building (low rise) was considered to be made energy efficient by placing the high thermal conductivity materials (Aluminum or silver pipes) along the inside face of the walls of the building rooms by extending their lower part to the sufficient ground contact where temperature is lower than the ambient room temperature.

2.0   NUMERICAL MODELING OF HEAT FLOW SYSTEM BETWEEN THE CONDUCTIVITY MEDIA AND THE UNDERGROUND SOIL MASS

This analysis is done to see the role of thermal conductivity media in cooling of the building. One room of a single story building is taken into consideration in numerical investigation of heat flow through the high thermal conductivity pipes into the underground soil. ANSYS 11 [16] general purpose finite element software was used to model the medias (conductivity media and underground soil) where heat flow was considered to be occurred. 8-noded linear thermal 3D solid elements (solid70) with one degree of freedom (DOF) at each node were used to discretize the entire systems. Temperature was defined at the air media inside of the room as well as underground soil. Temperature variation was considered within the underground soil of depth of 10 feet (according to Figure 2). In this depth temperatures were varied as 80.6^0F (27^0C) at zero depth, 75.2^0F (24^0C) at 3 feet depth, 71.6^0F (22^0C) at 6 feet depth and 68^0F (20^0C) at 10 feet depth. Thereafter, a constant temperature of 68^0F ($20\ ^0$C) was assumed in higher depth (>10 feet) of underground soil. Pipe temperature was considered as same as room temperature. The trend of this temperature variation was obtained from the study of Overby [15] (see Figure 3). Overby carried out research on environment which is dominated by moderate temperature. However, in this study investigation was performed for the tropical countries like Malaysia where temperatures vary 26^0C to 32^0C in day time. From the literature it is seen that room temperature increases vertically from floor to ceiling along interior (center) of the room. A vertical variation (80.6^0F at 0 height, 82.4^0F at height 4 feet and 85.1^0F at height 7.5 feet and above) of room air temperature between room height of 0 ft and 7.5 ft was considered in this study. A square cross sectional hollow aluminum pipes (1 inch by 1 inch outer dimension with thickness of 0.125 inch) were taken into consideration for the simplicity of the analysis. At first single pipe with underground soil media was analyzed to see the heat flow of the room through the pipe. Since temperature around the pipe (on four sides of the pipes) is considered to be the same as air temperature of the room and material and geometry of the pipe all through the pipe length are also same therefore, only one quarter of the pipe (quarter symmetry advantage) is taken into consideration in modeling. The entire heat flow problem was considered as conduction and convection problem of heat transfer system. Conductivity/convection properties of different materials are given in Table 1. Steady state condition of heat flow into the system was taken into consideration in this study.

Table 1. Conductivity/convection properties of different materials [17]

Name of Materials	Conductivity Coefficient (Btu/in.hr.^{0}F)	Film Coefficient (Btu/in^2.hr.^{0}F)
Aluminum	12.05	-
Concrete	0.03	
Soil (saturated sandy)	0.05	-
Air	-	0.012

3.0   COMBINED EFFECT OF MECHANICAL COOLING SYSTEM AND CONDUCTIVITY MEDIA ON THE COOLING SYSTEM OF THE BUILDING PARAGRAPH

In order to determine the combined effect of mechanical cooling system and conductivity media on the cooling system of the building, two cases were studied. In the first case, a 5 inch thick and 120 inch (height) by 62 inch (width) cement concrete (C.C) wall with only mechanical cooling system was simulated and analyzed for the outside temperature (due to sun radiation) and cooling set point temperature of the room and in the 2nd case the same system was modeled and analyzed in the presence of underground soil and conductivity media. In this case, 10 feet depth of soil with variation of temperature was taken into consideration. Chiller cooling system with the cooling set point temperature of 75.2^0F (24^0C) was considered as mechanical cooling system in this study. This system was simulated through the removal of heat flux from the interior face of the wall for a temperature difference of 10.8^0F (6^0C). This technique was also applied in simulating the variable temperature of underground soil. Heat flux was calculated using the heat flux and the temperature gradient relationship ($q = k \, \Delta T/d$, where q = heat flux, in Btu/in^2, k = thermal conductivity coefficient, ΔT = temperature difference, d = thickness of the media).

Outside air temperature due to sun radiation was considered as 86^0F (30^0C). Square cross sectional pipes were considered to be fixed with the inner surface of the C.C. wall at a spacing of 6 inch c/c. It was assumed that pipes temperature will be the same as the room temperature. The entire heat flow problem was considered as conduction and convection problems of heat transfer system. Thermal 3D solid (solid70) elements were used to discretize the entire system. Temperatures of the different mass medias (outside air temperature of the room and underground soil mass) were considered as the thermal load of the problem. Finite element discretization of the entire system is shown in Figure 4. The entire analysis was carried out for steady state condition of heat transfer problem.

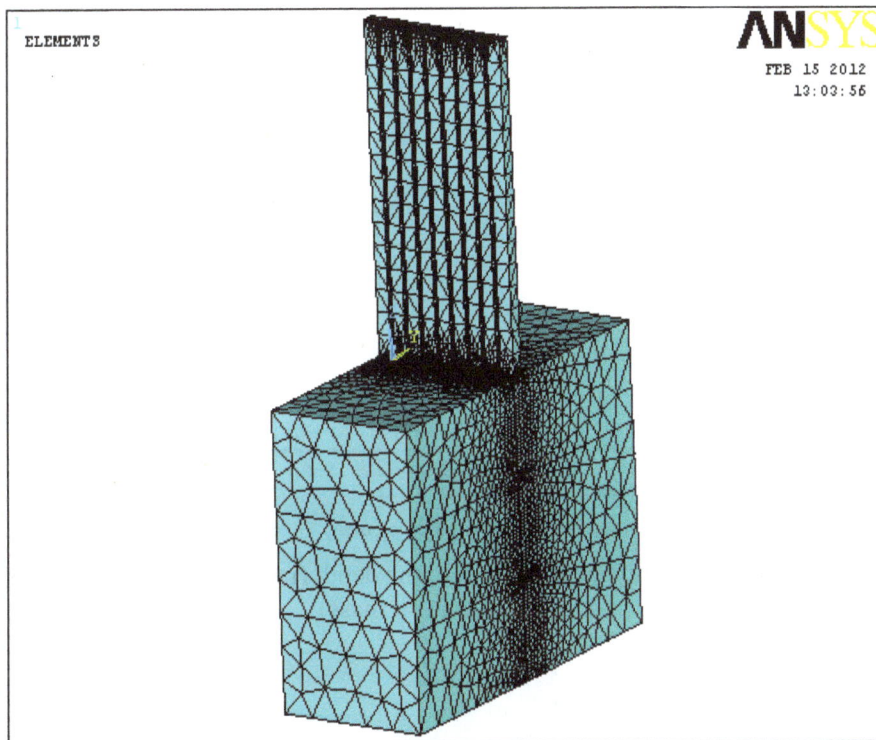

Figure 4: Finite element discretization

4.0   RESULTS AND DISCUSSIONS

ANSYS post processor was used to determine the temperature of the entire domain as well as its critical parts such as CC wall, pipes, etc. Variation of temperature along the depth of the pipe is shown in Figure 5 for room condition modeled without any mechanical cooling system. It is seen that temperatures decrease with the increased depth of the thermal conductivity pipe and minimum temperature is found at the region where soil temperature is constant. Contour plot of temperature of the room wall is shown in Figure 6 for room condition modeled with only mechanical cooling system. It is seen that temperature does not varied along the height of the wall. Variation of temperature along the depth of the pipe is shown in Figure 7 for room condition modeled with thermal conductivity pipes and mechanical cooling system. From this figure it is seen that temperature decreases with the depth of the pipe and along the inner surface of the wall. It is also seen that room temperature at the inner surface of the wall for this case is lower than those obtained for room condition modeled with mechanical cooling system alone (maximum temperature on the inner surface of the CC wall with thermal conductivity pipes and mechanical cooling system is   23.1 ^{0}C (73.58^{0}F) on the other hand inner surface temperature of the wall is 24^{0}C (75.20^{0}F) (from Figure 6) for CC wall with mechanical cooling system along. Due to fixing thermal conductivity pipes to the inner surface of the room wall in combination with mechanical cooling system, room temperature decreases by 1.7^{0}C (3^{0}F) (average value of temperature difference of pipe at top (0 inch depth -1.6^{0}C (2.86^{0}F)), at 45 inch depth (0.9^{0}C (1.62^{0}F)) and ground level (120 inch- 2.68^{0}C (4.82^{0}F)). Transverse temperature distribution along the width of the room wall is shown in Figure 8 for CC wall with thermal conductivity pipes and the mechanical cooling system. It is seen that pipe temperature is lower than the exterior wall temperature for room condition modeled with thermal conductivity pipe alone. In case of room conditioned modeled with thermal conductivity pipes and mechanical cooling system together it is seen that thermal conductivity pipe temperature is lower than the temperature of the inner surface of CC wall. This indicates that heat is flowing to the underground soil through the thermal conductivity pipes.

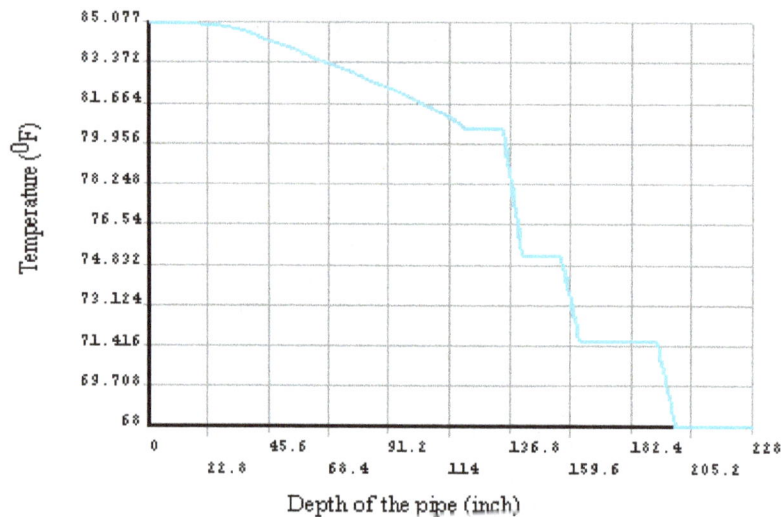

Figure 5: Variation of temperature (^{0}F) with the depth (inch) of pipe (single pipe and underground soil mass)

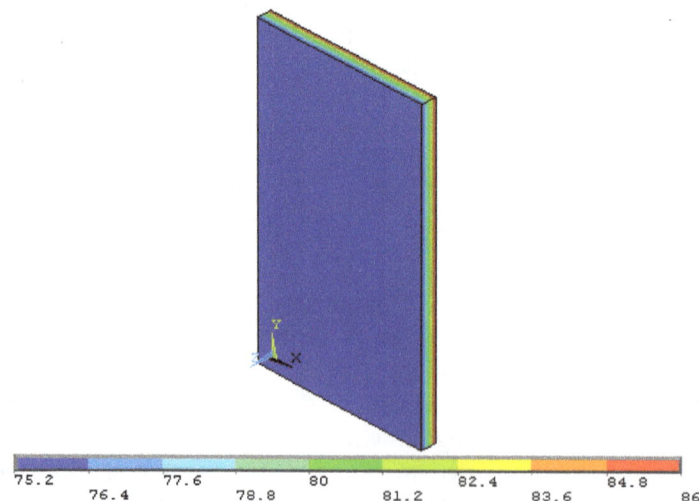

Figure 6: Contour plot of temperature (^{0}F) of CC wall with mechanical cooling system alone

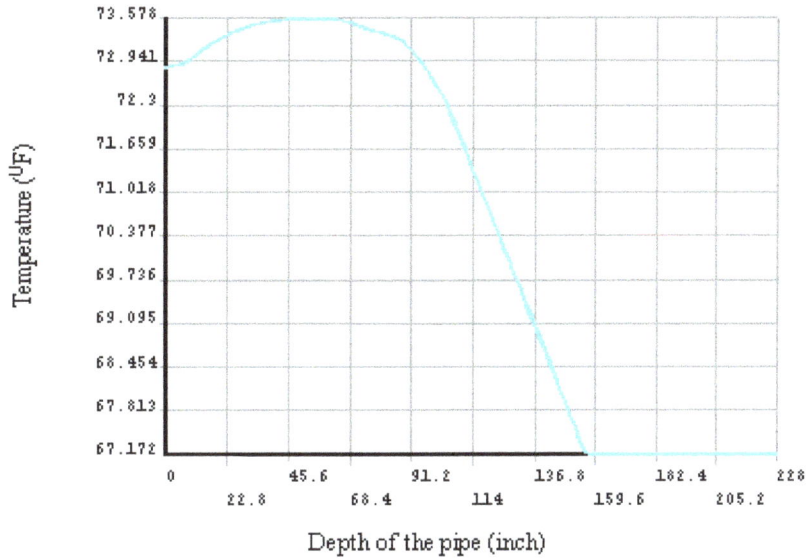

Figure 7: Variation of temperature (^{0}F) with the depth (inch) of pipe (CC wall with thermal conductivity pipes and mechanical cooling system)

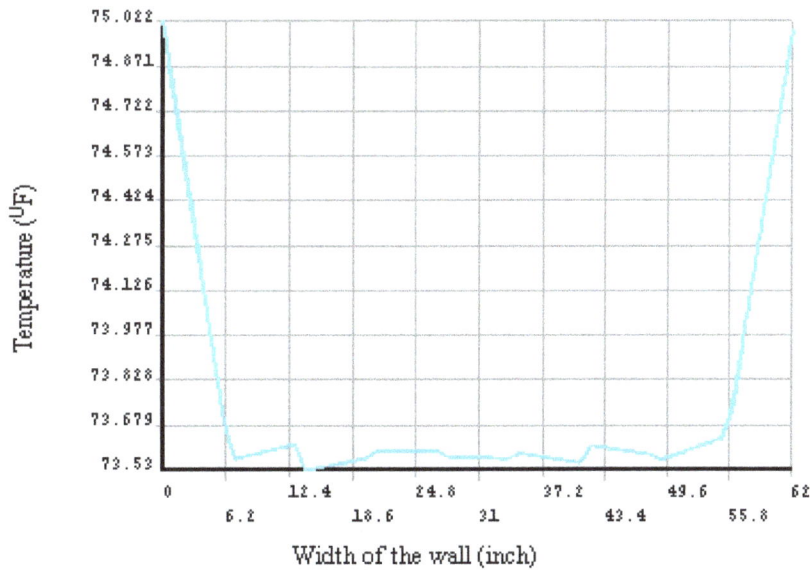

Figure 8: Temperature (^{0}F) distribution (transverse direction) along the interior side of the wall (CC wall with thermal conductivity pipes and mechanical cooling system)

Heat flux is the parameter that determines the heat flow per unit area of the body. Therefore, heat flux of the proposed cooling system was also determined at different locations of the entire domain. Heat flux vector plot along the pipe and underground soil mass (case I) is shown in Figure 9. From this figure it is seen that heat flows from higher temperature zone to the lower temperature zone. When heat flows from higher temperature zone to lower temperature zone, higher temperature zone starts to release heat and subsequently room temperature decreases. Heat flux variation along the depth of the pipe for single pipe and underground soil mass system is shown in Figure 10. From this figure it is seen that heat flux value is maximum in the soil region where temperature starts to vary (from top of the underground soil to a depth of 0.9 m (3 ft) inside the underground soil media).

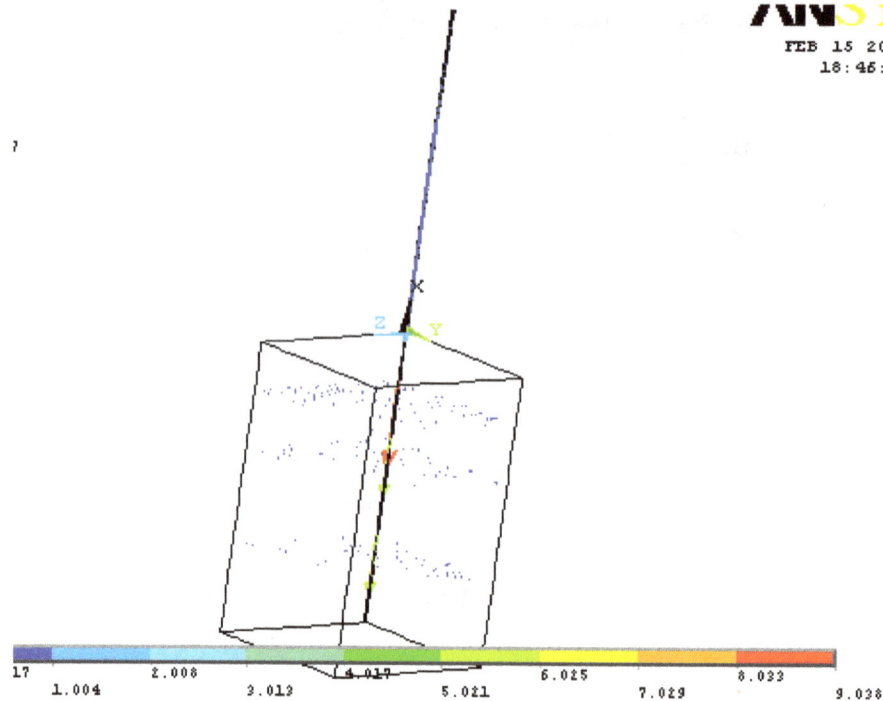

Figure 9: Vector plot of heat flux (single pipe and underground soil mass)

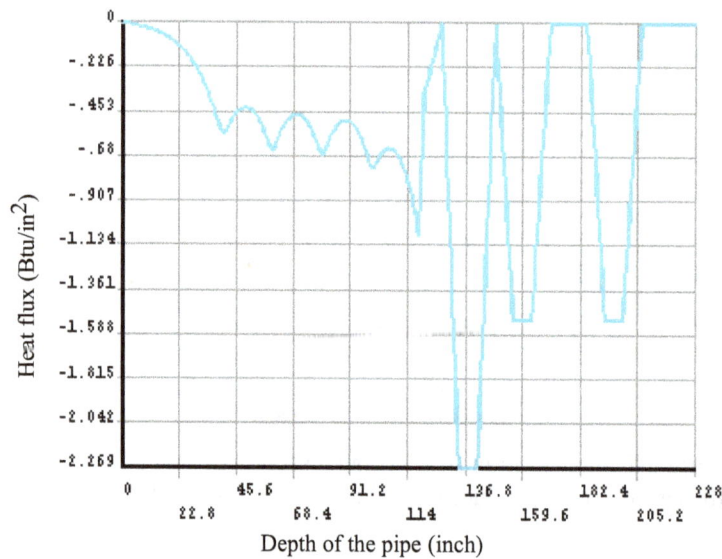

Figure 10: Variation of heat flux (Btu/in^2) with the depth of pipe (single pipe and soil mass)

When mechanical cooling system and thermal conductivity pipes were considered together in cooling system of the building room, heat flux variation was obtained much less than those obtained from the room condition modeled with thermal conductivity pipes alone. In this case significant temperature drop inside the room was obtained as illustrated in Figure 7.

5.0 CONCLUSIONS

From the numerical study carried out on heat transfer problem of different systems required for cooling residential/commercial low rise building it can be concluded that inside room temperature of the building could be possible to reduce if rooms are constructed with high thermal conductivity media (pipes) in connection with the underground soil where temperature remains constant and less than the ambient room temperature. From the analysis it is seen that square shaped aluminum pipes are very effective in transferring heat from the interior of the room to the underground soil and underground soil acts as cooling reservoir for the building built in tropical environment. A temperature drop of 2.8^0C (5^0F) between top and ground level heights of the square shaped thermal conductivity pipes was found in this study. Heat flux was obtained more at a depth of 1 foot of the underground soil and its direction was obtained downward. This result gives an indication of heat flow through the pipes from inside of the room to the underground soil. When the thermal conductivity pipes are considered to be added to the mechanical cooling system of the building, higher cooling set point temperature could be fixed up for the cooling of the room of the building as conductivity pipes share to transfer the heat from room to the underground soil. If the thermal conductivity pipes and the mechanical cooling system are incorporated together higher cooling set point of mechanical cooling system could make building room comfortable and the energy required making cool of the building through mechanical cooling could take much less.

ACKNOWLEDGEMENT

This study was carried out under the funding of Faculty of Engineering and Built Environment, Universiti Kebangsaan, Malaysia (UKM), Bangi, Selangor D.E., Malaysia

REFERENCES

[1] Kasuda. T and Archenbach. P.R (1965), "Earth Temperature amd Thermal Diffusivity at Selected Stations in the United States", ASHRAE Transactions, Vol. 71 (1)

[2] Crawford. C.B and Legget. R.F (1937), "Ground temperature investigations in Canada", Research paper no. 33 of the division of building research, Reprinted from the engineering journal, Vol 40 (3)

[3] Florides. G and Kalogirou. S (2004), "Annual ground temperature measurement at various depths, Higher Technical Institute, P.O. Box 20423, Nicosia 2152, Cyprus (ktisis.cut.ac.cy/bitstream/10488 /870/1/C55-PRT020-SET3.pdf)

[4] Nassar. Y, ElNoaman. A, Abutaima. A, Yousif. S and Salem. A (2006), "Evaluation of the underground soil thermal storage properties in Libya", Renewable Energy, Vol 3 (5), 2006, pages: 593-598.

[5] Bansal. N.K, Sodha. M.S, Bharadwaj. S.S (1983), "Performance of earth air tunnels", Energy Research, Vol.7, New Delhi: Wiley; pages: 156

[6] Nik. A. R, Kasran. B and Hassan. A (1986), "Soil Temperature Regime under Mixed Dipterocarp Forests of Peninsular Malysia", Partanika, Vol 9 (3), pages: 277-284

[7] Cui. W., Liao. Q., Chang. G., Chen. G., Peng. Q. and Jen. T.C. (2011), "Measurement and prediction of undisturbed underground temperature distribution", Proceedings of the ASME 2011 International Mechanical Engineering Congress & Exposition IMECE2011, Denver, Colorado, USA.

[8] Reimann. G, Boswell. H and Bacon. S (2007), "Ground cooling of ventilation air for energy efficient house in Malaysia: a case study of the cooltek house", Conference on Sustainable Building, Southeast Asia.

[9] Culver. G and Lund. J. W (1999), "Down hole heat exchanger", Geo-heat Center, GHC Bulletin, pages: 1-11

[10] Lund. J.W (2007), A report on heat exchanger, Geo-heat Center, Oregon Institute of Technology

[11] BINE informationsdienst (information, service, energy, expertise), project info 07/10, Project # 0327364A,B, ISSN 0937-8367, FIZ Karlsruke Publisher,76344 Egzenstein-Leopoldshafen, Germany

[12] Simone. A and Rode. C (2009), "Temperature distribution in riso flexhouse room with different heating control principle", Technical report, Department of Civil Engineering, DTU Civil Engineering-Report SR-09-07 (UK)

[13] Yasui. S, Yamanaka. T, Sagara. K., Kotani. H, Momoi. Y and Yamada. J. (2010), "Indoor thermal environment and vertical temperature gradient in large workshop of school without air conditioning", Report, Dept. of Architectural Eng., Osaka University, Suite 565-087, Osaka, Webpage link: server-labo4.arch.eng.osaka-u.ac.jp/~serverlabo4/.../10-i-16.pdf

 [14] Pollard. A, O'Driscoll. R and Pinder. D.N (2001), "The impact of Solar Radiation on the Air Temperature within a residential building", International Solar Energy Society 2001, Solar World Congress, Adelaide.

[15] Overby. H. (1994), "Measurement and calculation of vertical temperature gradients in rooms with convective flows", ROOMVENT 94, 4th International Conference on Air Distribution in Rooms, Cracow, Poland.

[16] ANSYS 11.0 (2007), ANSYS Inc., South Point, Canonsburg, U.S.A

[17] http://www.engineeringtoolbox.com/thermal-conductivity-d_429.html (search date, January 31, 2011)

11

SUITABILITY OF INDIAN HOT- ROLLED PARALLEL FLANGE SECTIONS FOR USE IN SEISMIC STEEL MOMENT RESISTING FRAMES

Kulkarni Swati Ajay[1], Vesmawala Gaurang R[2]

[1]Applied Mechanics Department, SVNIT, Surat, India
[2]Applied Mechanics Department, SVNIT, Surat, India

*Corresponding E-mail : swatiakulkarni@gmail.com

ABSTRACT

Use of parallel flange I beam sections is advantageous than tapered flange I beam sections due to, increased lateral stiffness, sections do not have sloping flanges and excessive material in web and easy to weld and bolt. Nowadays the hot rolled parallel flange, narrow parallel flange beams (NPB) and wide parallel flange beams (WPB) sections as per Indian standards, having yield stress, 300 MPa, 350 MPa and 410 MPa are being manufactured. Available range of these sections can be used for steel moment resisting frames (SMRF's) and prequalified connections as per AISC codes. When the cross section of a steel shape is subjected to large compressive stresses, the thin plates that make up the cross section may buckle before the full strength of the member is attained if the thin plates are too slender. This failure mode may be prevented by selecting suitable width-to-thickness ratios of component plates. In the present exercise, a suitability of NPB and WPB section for use in SMRF's as per width-to-thickness limitations of AISC 341-2010 and AISC 341-2005 codal provisions is studied.

Keywords: *steel moment resisting frames, local buckling, slenderness limits, parallel flange, I beam*

1.0 INTRODUCTION

Steel moment resisting frame structures are frequently used as seismic load resisting systems for building in seismic regions. SMRF's are rectilinear assemblies of columns and beams that are typically joined by welding or high-strength bolting or both. Resistance to lateral load is provided by flexural and shearing actions in the beams and columns. Lateral stiffness is provided by flexural stiffness of the beams and columns [1]. AISC- Seismic provisions for structural steel buildings [2-4], defines three types of seismic steel moment resisting frames: special moment frames (SMF), intermediate moment frames (IMF) and ordinary moment frames (OMF). When subjected to the forces resulting from the motions of the design earthquake: SMF are expected to withstand significant inelastic deformations; IMF are expected to withstand limited inelastic deformations; OMF are expected to withstand minimal inelastic deformations, in their members and connections. Reliable inelastic deformation requires that width-to-thickness ratios of compression elements be limited to a range that provides a cross section resistant to local buckling into the inelastic range [5]. Seismic provisions, require member flanges to be continuously connected to the web(s) and width-thickness ratios of the compression elements must be less than or equal to those that are resistant to local buckling when stressed into the inelastic range.

Based on the width-to-thickness ratios ($\frac{b_f}{t_f}$, $\frac{h}{t_w}$) of the plate elements that make up

member's cross sections, AISC 341-2005 [3] & 360-2005[6] uses a term 'seismically compact',

'compact' and 'noncompact' to categorize the members. However, as per AISC 341-2010 [2], members are classified into 'highly ductile' and 'moderately ductile'.

Compact section is a section capable of developing a fully plastic stress distribution and possessing a rotation capacity of approximately three before the onset of local buckling Noncompact section is a section that can develop the yield stress in its compression elements before local buckling occurs, but cannot develop a rotation capacity of three. A higher level of compactness termed as 'seismically compact' is a section, expected to be able to achieve a level of deformation ductility of at least 4. Beam and column members should satisfy the requirements of width-to-thickness ratios as specified by AISC codes for above members, unless otherwise qualified by tests.

Highly ductile member is a member expected to undergo significant plastic rotation (more than 0.02 rad) from either flexure or flexural buckling under the design earthquake. Moderately ductile member is a member expected to undergo moderate plastic rotation (0.02rad or less) from either flexure or flexural buckling under the design earthquake. Beam and column members used for SMF are highly ductile members, for IMF moderately ductile members and for OMF there are no limitations on width-to-thickness ratios of members.

Usually, as per Indian standard (IS), IS 800-2007, IS 808 -1989 and IS 1852 -1985 [7-9] hot rolled tapered flange (Fig. 1A), I- sections are classified into four types namely, light (ISLB), medium (ISMB), wide flange (ISWB) and heavy (ISHB) beams. Further, as per IS 12778 -2004, IS12779-1989 [10, 11] hot rolled parallel flange sections are classified as (Fig. 1B) as narrow parallel flange beams (NPB), wide parallel flange beams (WPB) and parallel flange bearing pile sections (PBP). All above sections having yield stress 250MPa [12] are most commonly produced and used for steel structures in India. Parametric analysis [13] has shown that Indian hot- rolled I sections (parallel as well as tapered) having yield stress 250 do not meet compactness requirements specified in Indian standards as well as those of countries with advanced seismic provision for SMF. Even those that satisfy the stability requirements, their sizes are so small that they are insufficient from strength and stiffness points of view to be able to construct large span and high rise earthquake resistant constructions in strong seismic regions.

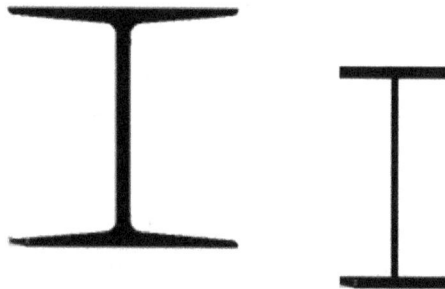

Figure 1: A) Hot rolled tapered flange I section, B) Hot rolled parallel flange I section

After the 1994 Northridge and 1995 Kobe earthquake, a significant amount of research activity was initiated on the behavior of fully restrained steel connections. Various types of beam-to-column connections have been proposed and investigated. So far, three general approaches were followed in improving connection details: 1) improving unreinforced connections / toughening schemes, 2) strengthening approach: strengthening connection by addition of cover plates, ribs or haunches, and 3) weakening approach: locally weakening the beam away from the column face by reduced beam section (RBS) or slotted web [14, 15, 16]. All these schemes are often used in combination. The AISC codes [2-4, 6, 15-17] specifies the guidelines about design of seismic steel moment resisting frames, beam-to-column connection details, width-to-thickness limitation for members and other details . Although these connections/schemes are widely investigated and used in US, Japan and Europe, however design of such type of connections are not presented and used in India.

Nowadays the hot rolled parallel flange NPB and WPB sections as per IS 8500 -1991[18], having yield stress, 300 MPa, 350 MPa and 410 MPa are being manufactured in India. Available range of these sections can be used for steel moment resisting frames (SMRF's) as well as prequalified connections as per AISC 358-2005, 2010 [15, 16]. As per IS code 12778 -2004 [10] NPB sections are mostly used as beams and WPB sections are generally used as beams or columns. In the present exercise, these parallel flange members are classified as per width-to-thickness limitation of AISC 341-2005 and AISC 341-2010 provisions and their suitability for use in seismic steel moment resisting frames is studied.

2.0   CLASSIFICATION OF SECTIONS FOR LOCAL BUCKLING

The cross sections of steel shapes tend to consist of an assembly of thin plates. When the cross section of a steel shape is subjected to large compressive stresses, the thin plates that make up the cross section, may buckle before the full strength of the member is attained, if the thin plates are too slender. When a cross sectional element fails in buckling, then the member capacity is reached. Consequently, local buckling becomes a limit state for the strength of steel shapes subjected to compressive stress. This failure mode may be prevented by selecting suitable width-to-thickness ratios (Fig. 2) of component plates [19].

In the following sections, classification of NPB (beam) and WPB (column) sections is considered as per width-to-thickness limitation of AISC 341-2005 and AISC 341-2010 provisions. Usually NPB sections are used for beams, slenderness check is also considered to verify their suitability as a column member.

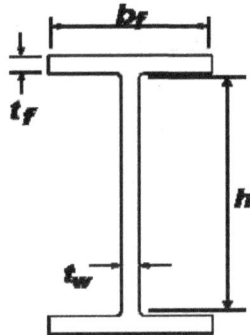

Figure 2: Typical parallel flange section

2.1   CLASSIFICATION ACCORDING TO AISC 341-2005

With respect to the following formulae, classification of NPB and WPB sections as per AISC 341-2005 & 360-2005 is as shown in Equations 1 to 7.
Limiting width-to-thickness ratios for compression elements for seismically compact members:

$$\frac{b_f}{2t_f} \leq 0.30 \sqrt{\frac{E}{F_y}} \tag{1}$$

$$C_a \leq 0.125, \frac{h}{t_w} \leq 3.14 \sqrt{\frac{E}{F_y}} \left[1 - 1.54 C_a\right] \tag{2}$$

$$C_a > 0.125, \frac{h}{t_w} \leq 1.12 \sqrt{\frac{E}{F_y}} \left[2.33 - C_a\right] \geq 1.49 \sqrt{\frac{E}{F_y}} \tag{3}$$

Where

$$C_a = \frac{P_u}{\Phi_b P_y} \quad ..... \text{ for} \qquad \text{Load and Resistance Factor Design (LRFD)}$$

$$C_a = \frac{\Omega_b P_a}{P_y} \quad ...... \text{ for} \qquad \text{Allowable Strength Design (ASD)}$$

Limiting width-to-thickness ratios for compression elements for compact members:

$$\frac{b_f}{2t_f} \leq 0.38\sqrt{\frac{E}{F_y}}$$

(4)

$$\frac{h}{t_w} \leq 3.76\sqrt{\frac{E}{F_y}}$$

(5)

Limiting width-to-thickness ratios for compression elements for noncompact members:

$$\frac{b_f}{2t_f} \leq 1\sqrt{\frac{E}{F_y}}$$

(6)

$$\frac{h}{t_w} \leq 5.70\sqrt{\frac{E}{F_y}}$$

(7)

As per IS code [10], 70 numbers of NPB and 122 numbers of WPB sections are available. Therefore, 192 numbers of member can be used as beam and 122 as column. Beam members (total number of sections) [20], satisfying $\frac{b_f}{2t_f}$ and $\frac{h}{t_w}$ ratio are as shown in Table 1.

Table 1: NPB and WPB sections satisfying width-to-thickness ratio as beam

Slenderness Limit		Steel with F_y MPa					
		300		350		410	
		NPB	WPB	NPB	WPB	NPB	WPB
SMF	$\frac{b_f}{2t_f} \leq 0.30\sqrt{\frac{E}{F_y}}$	≤ 7.74		≤ 7.17		≤ 6.63	
		56	73	51	64	38	56
	$\frac{h}{t_w} \leq 3.14\sqrt{\frac{E}{F_y}}$	≤ 81.07		≤ 75.06		≤ 69.35	
		ALL	ALL	ALL	ALL	ALL	ALL
IMF	$\frac{b_f}{2t_f} \leq 0.38\sqrt{\frac{E}{F_y}}$	≤ 9.80		≤ 9.08		≤ 8.93	
		69	96	68	87	67	81
	$\frac{h}{t_w} \leq 3.76\sqrt{\frac{E}{F_y}}$	≤ 97.08		≤ 89.88		≤ 83.04	
		ALL	ALL	ALL	ALL	ALL	ALL

WPB members are considered as column members. Table 2 Shows: a) Column sections satisfying $\frac{b_f}{2t_f}$ limit as recommended by Equations 1, b) Maximum web depth-to-thickness ratio is given by Equation 2 and 3, giving a range of 38.47, 35.59 and 32.91 for $\frac{P_u}{\phi P_y} = 1$ and increasing it to 81.07, 75.06 and 69.35 for $P_u = 0$, for steel having yield stress 300, 350 and 410MPa respectively. Column sections (total number of sections) used for IMF satisfying limits as per Equations 4 & 5 are shown in Table 3.

Table 2: WPB sections satisfying width-to-thickness as column

Sl. No.	Section number as per IS code [10]	Section WPB	SMF Yield Stress (MPa)					
			300		350		410	
			$\dfrac{b_f}{2t_f}$	$\dfrac{h}{t_w}$	$\dfrac{b_f}{2t_f}$	$\dfrac{h}{t_w}$	$\dfrac{b_f}{2t_f}$	$\dfrac{h}{t_w}$
			$\leq$ 7.74	Range, 38.47 for $\dfrac{P_u}{\Phi_c P_y}=1$ and increasing to 81.07 for $P_u=0$	$\leq$ 7.17	Range, 35.59 for $\dfrac{P_u}{\Phi_c P_y}=1$ and increasing to75.06 for $P_u=0$	$\leq$ 6.63	Range, 32.91 for $\dfrac{P_u}{\Phi_c P_y}=1$ and increasing to 69.35 for $P_u=0$
1	96	540×300×12.5×24						√
2	100	590×300×13×25			√			√
3	104	640×300×13.5×26	√		√			√
4	105	650×300×16×31						√
5	108	690×300×14.5×27	√		√			√
6	109	700×300×17×32						√
7	112	790×300×15×28	√		√			√
8	113	800×300×17.5×33	√		√			√
9	116	840×292×14.7×21.3	√		√			
10	117	846×293×15.4×24.3	√		√			√
11	118	851×294×16.1×26.8	√		√			√
12	119	859×292×17×30.8	√		√			
13	120	870×300×15×20	√					
14	121	890×300×16×30	√		√			√
15	122	900×292×18.5×35	√		√			√
Total			11		11		13	

Table 3: WPB sections satisfying width-to-thickness as column

Slenderness Limit		Steel with F_y MPa		
		300	350	410
		WPB	WPB	WPB
IMF	$\dfrac{b_f}{2t_f} \leq 0.38\sqrt{\dfrac{E}{F_y}}$	≤9.80	≤9.08	≤8.93
		96	87	81
	$\dfrac{h}{t_w} \leq 3.76\sqrt{\dfrac{E}{F_y}}$	≤97.08	≤89.88	≤83.04
		ALL	ALL	ALL	ALL	ALL	ALL

Mostly all WPB and NPB sections satisfy limit for OMF.

2.2 CLASSIFICATION ACCORDING TO AISC 341-2010

Classification of NPB and WPB sections as highly ductile and moderately ductile according to following formulae as per AISC341-2010 is shown in Equations 8 to 13.

Limiting width-to-thickness ratios for compression elements for highly ductile members:

$$\frac{b_f}{2t_f} \le 0.30 \sqrt{\frac{E}{F_y}} \tag{8}$$

$$C_a \le 0.125, \frac{h}{t_w} \le 2.45 \sqrt{\frac{E}{F_y}}\left[1 - 0.93 C_a\right] \tag{9}$$

$$C_a > 0.125, \frac{h}{t_w} \le 0.77 \sqrt{\frac{E}{F_y}}\left[2.93 - C_a\right] \ge 1.49 \sqrt{\frac{E}{F_y}} \tag{10}$$

Limiting width-to-thickness ratios for compression elements for moderately ductile members:

$$\frac{b_f}{2t_f} \le 0.38 \sqrt{\frac{E}{F_y}} \tag{11}$$

$$C_a \le 0.125, \frac{h}{t_w} \le 3.76 \sqrt{\frac{E}{F_y}}\left[1 - 2.75 C_a\right] \tag{12}$$

$$C_a > 0.125, \frac{h}{t_w} \le 1.12 \sqrt{\frac{E}{F_y}}\left[2.33 - C_a\right] \ge 1.49 \sqrt{\frac{E}{F_y}} \tag{13}$$

Where

$$C_a = \frac{P_u}{\Phi_c P_y} \quad \ldots\ldots \text{ for } \qquad \text{Load and Resistance Factor Design (LRFD)}$$

$$C_a = \frac{\Omega_c P_a}{P_y} \quad \ldots\ldots \text{ for } \qquad \text{Allowable Strength Design (ASD)}$$

For I-shaped beams in SMF systems, where C_a is less than or equal to 0.125, the limiting ratio $\frac{h}{t_w}$ shall not exceed $2.45 \sqrt{\frac{E}{F_y}}$. For I-shaped beams in IMF systems, where C_a is less than or equal to 0.125, the limiting width-to-thickness ratio shall not exceed $3.76 \sqrt{\frac{E}{F_y}}$. Beam members (total number of sections), satisfying $\frac{b_f}{2t_f}$ and $\frac{h}{t_w}$ ratio are as shown in Table 4.

Table 4: NPB and WPB sections satisfying width-to-thickness ratio as beam

Slenderness Limit		Steel with F_y MPa					
		300		350		410	
		NPB	WPB	NPB	WPB	NPB	WPB
SMF	$\frac{b_f}{2t_f} \le 0.30 \sqrt{\frac{E}{F_y}}$	≤7.74		≤7.17		≤6.63	
		56	73	51	64	38	56
	$\frac{h}{t_w} \le 2.45 \sqrt{\frac{E}{F_y}}$	≤63.25		≤58.56		≤54.11	
		ALL	ALL	ALL	ALL	ALL	ALL
IMF	$\frac{b_f}{2t_f} \le 0.38 \sqrt{\frac{E}{F_y}}$	≤9.80		≤9.08		≤8.93	
		69	96	68	87	67	81
	$\frac{h}{t_w} \le 3.76 \sqrt{\frac{E}{F_y}}$	≤97.08		≤89.88		≤83.04	
		ALL	ALL	ALL	ALL	ALL	ALL

WPB sections satisfying $\frac{b_f}{2t_f}$, $\frac{h}{t_w}$ limt for SMF and IMF as recommended by equations 8, 9, 10, 11, 12 and 13 are shown in Table 5 & 6.

Table 5: WPB sections satisfying width-to-thickness ratio as column

Sl. No.	Section number as per IS code [10]	Section WPB	SMF					
			Yield Stress (MPa)					
			300		350		410	
			$\frac{b_f}{2t_f}$	$\frac{h}{t_w}$	$\frac{b_f}{2t_f}$	$\frac{h}{t_w}$	$\frac{b_f}{2t_f}$	$\frac{h}{t_w}$
			≤ 7.74	Range, 38.47 for $\frac{P_u}{\Phi_c P_y}=1$ and increasing to 63.25 for $P_u = 0$	≤ 7.17	Range, 35.61 for $\frac{P_u}{\Phi_c P_y}=1$ and increasing to 58.56 for $P_u = 0$	≤ 6.63	Range, 32.91 for $\frac{P_u}{\Phi_c P_y}=1$ and increasing to 54.11 for $P_u = 0$
1	96	540×300×12.5×24						√
2	100	590×300×13×25				√		√
3	104	640×300×13.5×26		√		√		√
4	105	650×300×16×31						√
5	108	690×300×14.5×27		√		√		√
6	109	700×300×17×32						√
7	112	790×300×15×28		√		√		√
8	113	800×300×17.5×33		√		√		√
9	116	840×292×14.7×21.3		√		√		
10	117	846×293×15.4×24.3		√		√		√
11	118	851×294×16.1×26.8		√		√		√
12	119	859×292×17×30.8		√		√		√
13	120	870×300×15×20		√				
14	121	890×300×16×30		√		√		√
15	122	900×292×18.5×35		√		√		√
Total				11		11		13

Table 6: WPB sections satisfying width-to-thickness ratio as column

Sl. No.	Section number as per IS code [10]	Section WPB	IMF Yield Stress (MPa)					
			300		350		410	
			$\frac{b_f}{2t_f}$	$\frac{h}{t_w}$	$\frac{b_f}{2t_f}$	$\frac{h}{t_w}$	$\frac{b_f}{2t_f}$	$\frac{h}{t_w}$
			≤ 9.80	Range, 38.47 for $\frac{P_u}{\Phi_c P_y}=1$ and increasing to 97.08 for $P_u=0$	≤ 9.08	Range, 35.61 for $\frac{P_u}{\Phi_c P_y}=1$ and increasing to 89.88 for $P_u=0$	≤ 8.93	Range, 32.91 for $\frac{P_u}{\Phi_c P_y}=1$ and increasing to 83.04 for $P_u=0$
1	91	480×300×11.5×18						√
2	96	540×300×12.5×24						√
3	99	571×300×12×15.5		√				
4	100	590×300×13×25				√		√
5	103	620×300×12.5×16		√				
6	104	640×300×13.5×26		√		√		√
7	105	650×300×16×31						√
8	107	670×300×13×17		√		√		
9	108	690×300×14.5×27		√		√		√
10	109	700×300×17×32						√
11	111	770×300×14×18		√		√		√
12	112	790×300×15×28		√		√		√
13	113	800×300×17.5×33		√		√		√
14	115	835×292×14×18.8		√		√		√
15	116	840×292×14.7×21.3		√		√		√
16	117	846×293×15.4×24.3		√		√		√
17	118	851×294×16.1×26.8		√		√		√
18	119	859×292×17×30.8		√		√		√
19	120	870×300×15×20		√		√		√
20	121	890×300×16×30		√		√		√
21	122	900×292×18.5×35		√		√		√
Total				16		15		18

Mostly all WPB and NPB sections can be used for OMF

2.3 CLASSIFICATION OF NPB SECTIONS CONSIDERING AS A COLUMN MEMBER

Though, WPB sections are considered as column member, it can be observed from above Tables 2, 3, 5 & 6: a) for SMF as per both codes, sections which qualify the limit are same and very few in number; b) for IMF according to AISC 341-2005 more WPB sections satisfy slenderness ratio than AISC 341-2010; c) mostly all WPB sections satisfy limit for OMF.

Therefore, width-to-thickness check is applied to NPB sections considering them as a column section as per both codes.

NPB sections satisfying $\dfrac{b_f}{2t_f}$, $\dfrac{h}{t_w}$ limit for SMF as recommended by AISC 341-2005, as per

Equations 1, 2, 3 are shown in Table 7. For IMF, NPB sections satisfying the limits are as per Table 1.

Table 7: NPB sections satisfying width-to-thickness ratio as column

Sl. No.	Section number as per IS code [10]	Section NPB	SMF					
			Yield stress (MPa)					
			300		350		410	
			$\dfrac{b_f}{2t_f}$	$\dfrac{h}{t_w}$	$\dfrac{b_f}{2t_f}$	$\dfrac{h}{t_w}$	$\dfrac{b_f}{2t_f}$	$\dfrac{h}{t_w}$
			$\leq$ 7.74	Range ,38.47 for $\dfrac{P_u}{\Phi_c P_y}=1$ and increasing to 81.07 for $P_u=0$	$\leq$ 7.17	Range, 35.59 for $\dfrac{P_u}{\Phi_c P_y}=1$ and increasing to 75 for $P_u=0$	$\leq$ 6.63	Range, 32.91for $\dfrac{P_u}{\Phi_c P_y}=1$ and increasing to 69.35 for $P_u=0$
1	29	270×135×6.6×10.2						√
2	35	313×166×6.6×11.2	√					
3	36	317×167×7.6×13.2						√
4	41	330×160×7.5×11.5			√			
5	43	357.6×170×6.6×11.5	√					
6	44	360×170×8×12.7			√			
7	47	397×180×7×12	√					
8	48	400×180×8.6×13.5	√		√			
9	49	404×182×9.7×15.5						√
10	50	400×200×8×13	√					
11	51	447×190×7.6×13.1	√					
12	52	450×190×9.4×14.6	√		√			√
13	53	456×192×11×17.6						√
14	54	497×200×8.4×14.5	√		√			
15	55	500×200×10.2×16	√		√			√
16	56	506×202×12×19						√
17	57	547×210×9×15.7	√		√			
18	58	550×210×11.1×17.2	√		√			√
19	59	556×212×12.7×20.2			√			√
20	60	597×220×9.8×17.5	√		√			√

21	61	600×220×12×19	√	√	√
22	62	610×224×15×24			√
23	64	695×250×11.5×16.5	√		
24	65	700×250×12.5×19	√	√	√
25	66	704×250×13×21	√	√	√
26	67	709×250×14.5×23.5	√	√	√
27	69	760×270×14.4×21.6	√	√	√
28	70	770×270×15.6×26.6	√	√	√
Total			19	16	17

NPB sections satisfying $\dfrac{b_f}{2t_f}$, $\dfrac{h}{t_w}$ limt for SMF and IMF as recommended by AISC 341-2010, as per Equations 8,9,10,11,12 and 13, are shown in Table 8 and 9 respectively.

Table 8: NPB sections satisfying width-to-thickness ratio as column

Sl. No.	Section number as per IS code [10]	Section NPB	SMF					
			Yield stress (MPa)					
			300		350		410	
			$\dfrac{b_f}{2t_f}$	$\dfrac{h}{t_w}$	$\dfrac{b_f}{2t_f}$	$\dfrac{h}{t_w}$	$\dfrac{b_f}{2t_f}$	$\dfrac{h}{t_w}$
			≤ 7.74	Range ,38.47 for $\dfrac{P_u}{\Phi_c P_y}=1$ and increasing to 63.25 for $P_u = 0$	≤ 7.17	Range, 35.61 for $\dfrac{P_u}{\Phi_c P_y}=1$ and increasing to 58.56 for $P_u = 0$	≤ 6.63	Range,32.91 for $\dfrac{P_u}{\Phi_c P_y}=1$ and increasing to 54.11 for $P_u = 0$
1	29	270×135×6.6×10.2						√
2	35	313×166×6.6×11.2	√					
3	36	317×167×7.6×13.2						√
4	41	330×160×7.5×11.5				√		
5	43	357.6×170×6.6×11.5	√					
6	44	360×170×8×12.7				√		
7	47	397×180×7×12	√					
8	48	400×180×8.6×13.5	√			√		
9	49	404×182×9.7×15.5						√
10	50	400×200×8×13	√					
11	51	447×190×7.6×13.1	√					
12	52	450×190×9.4×14.6	√			√		√
13	53	456×192×11×17.6						√
14	54	497×200×8.4×14.5	√			√		
15	55	500×200×10.2×16	√			√		√
16	56	506×202×12×19						√
17	57	547×210×9×15.7	√			√		
18	58	550×210×11.1×17.2	√			√		√

19	59	556×212×12.7×20.2		√	√
20	60	597×220×9.8×17.5	√	√	√
21	61	600×220×12×19	√	√	√
22	62	610×224×15×24			√
23	64	695×250×11.5×16.5	√		
24	65	700×250×12.5×19	√	√	√
25	66	704×250×13×21	√	√	√
26	67	709×250×14.5×23.5	√	√	√
27	69	760×270×14.4×21.6	√	√	√
28	70	770×270×15.6×26.6	√	√	√
Total			19	16	17

Table 9: NPB sections satisfying width-to-thickness ratio as column

Sl. No.	Section number as per IS code [10]	Section NPB	IMF					
			Yield stress (MPa)					
			300		350		410	
			$\dfrac{b_f}{2t_f}$	$\dfrac{h}{t_w}$	$\dfrac{b_f}{2t_f}$	$\dfrac{h}{t_w}$	$\dfrac{b_f}{2t_f}$	$\dfrac{h}{t_w}$
			≤ 9.80	Range, 38.47 for and $\dfrac{P_u}{\Phi_c P_y}=1$ increasing to 97.08 for $P_u=0$	≤ 9.08	Range, 35.61 for and $\dfrac{P_u}{\Phi_c P_y}=1$ increasing to 89.36 for $P_u=0$	≤ 8.93	Range, 32.91 for and $\dfrac{P_u}{\Phi_c P_y}=1$ increasing to 83.04 for $P_u=0$
1	5	177×91×4.3×6.5						√
2	8	197×100×4.5×7						√
3	17	217×110×5×7.7						√
4	20	237×120×5.2×8.3				√		√
5	23	250×125×6×9						√
6	24	258×146×6.1×9.2						√
7	28	267×135×5.5×8.7		√		√		√
8	29	270×135×6.6×10.2						√
9	31	297×150×6.1×9.2		√		√		
10	32	300×150×7.1×10.7						√
11	34	310×165×5.8×9.7		√		√		
12	35	313×166×6.6×11.2		√		√		√
13	36	317×167×7.6×13.2						√
14	40	327×160×6.5×10		√		√		√
15	41	330×160×7.5×11.5				√		√
16	43	357.6×170×6.6×11.5		√				√
17	44	360×170×8×12.7				√		√
18	45	364×172×9.2×14.7						√
19	47	397×180×7×12		√		√		√
20	48	400×180×8.6×13.5		√		√		√
22	49	404×182×9.7×15.5						√
23	50	400×200×8×13		√		√		√
24	51	447×190×7.6×13.1		√		√		√
25	52	450×190×9.4×14.6		√		√		√
26	53	456×192×11×17.6						√

27	54	497×200×8.4×14.5	√	√	√
28	55	500×200×10.2×16	√	√	√
29	56	506×202×12×19			√
30	57	547×210×9×15.7	√	√	√
31	58	550×210×11.1×17.2	√	√	√
32	59	556×212×12.7×20.2		√	√
33	60	597×220×9.8×17.5	√	√	√
34	61	600×220×12×19	√	√	√
35	62	610×224×15×24			√
36	63	694×250×9×16	√	√	√
37	64	695×250×11.5×16.5	√	√	√
38	65	700×250×12.5×19	√	√	√
39	66	704×250×13×21	√	√	√
40	67	709×250×14.5×23.5	√	√	√
41	68	750×265×13.2×16.6	√	√	√
42	69	760×270×15.6×26.6	√	√	√
Total			24	28	40

If NPB sections are considered as column member it can be observed from above Tables 1, 7, 8 and 9: a) for SMF as per both codes, sections which qualify the limit are same and reasonable in number; b) for IMF according to AISC 341-2005 more NPB sections satisfy slenderness ratio than AISC 341-2010; c) mostly all NPB sections satisfy limit for OMF.

In the above paper, adequacy of Indian hot rolled parallel flange sections having yield stress 300MPa, 350 MPa and 410 MPa, as per slenderness limits of AISC 341-2005 & AISC 341-2010 code is discussed. Thus, suitable parallel flange section of Indian profile, satisfying width-to-thickness limit can be selected from the above Tables to build planned SMRF i.e. SMF, IMF, OMF and prequalified connection.

3.0　CONCLUSIONS

An exercise was carried out to understand suitability of parallel flange sections (Indian profile) for use in seismic steel moment resisting frames, reflects following:

- Guidelines provided in tabular form to choose sections for proposed moment frame.
- Sufficient numbers of beam sections are available for SMF, IMF and OMF. Column sections (WPB) suitable for SMF are available in limited number.
- Slenderness check applied to NPB members shows that, these sections, which satisfy the slenderness limit can be used as column members, however, when both (NPB and WPB) sections are considered, more choice is available for use as column members for SMF and IMF.
- When used as a column member for IMF, according to AISC 341-2005 more NPB & WPB sections satisfy slenderness ratio than AISC 341-2010.
- With the available members, suitability of these sections for connections with cover plates, ribs, haunches, reduced beam sections, slotted web etc. can be studied for Indian parallel flange profile sections.

REFERENCES

[1] Farzad Naeim (2001), "Seismic Design Handbook", Kluwer Acedemic Publishers Group 2nd edition, USA.

[2] ANSI/AISC 341-10 (2010), "Seismic provisions for structural steel building", American Institute of Steel Construction, Chicago, IL.

[3] ANSI/AISC 341-05 (2005), "Seismic provisions for structural steel buildings- including supplement No 1", American Institute of Steel Construction, Chicago, IL.

[4] ANSI/AISC 341-02 (2002), "Seismic provisions for structural steel buildings", American Institute of Steel Construction, Chicago, IL.

[5] Hamburger RO, Krawinlker H, Malley JM, Adan SM (2009), "Seismic Design of Steel Special Moment Frame- A Guide for Practicing Engineers", NEHRP Seismic Design Technical Brief No. 2, USA.

[6] ANSI/AISC 360-05 (2005), "Specification for structural steel buildings", American Institute of Steel Construction, Chicago, IL.

[7] IS-800 (2007), "General construction in steel- code of practice", Bureau of Indian Standards, New Delhi.

[8] IS-808 (1989), "Dimensions for hot rolled steel beam, column, channel and angle sections", Bureau of Indian Standards, New Delhi.

[9] IS-1852 (1985), "Specification for rolling and cutting tolerances for hot-rolled steel products", Bureau of Indian Standards, New Delhi.

[10] IS-12778 (2004), "Hot rolled parallel flange steel sections for beams, columns and bearing piles- dimensions and section properties", Bureau of Indian Standards, New Delhi.

[11] IS-12779 (1989), "Rolling and cutting tolerences for hot rolled parallel flange beams and columns section – Specifications", Bureau of Indian Standards, New Delhi.

[12] IS-2062 (1999), "Steel for general structural purposes- specification", Bureau of Indian Standards, New Delhi.

[13] Goswami R, Arlekar JN, Murthy CVR (2006), "Limitations of available Indian hot-rolled I-sections for use in seismic steel MRFs", Report nicee, IIT Kanpur.

[14] Subramanian N (2008), "Design of Steel Structures", Published in India by Oxford University Press.

[15] ANSI/AISC 358-10 (2010), "Prequalified connections for special and intermediate steel moment frames for seismic applications – Including supplement No. 1", American Institute of Steel Construction, Chicago, IL.

[16] ANSI/AISC 358-05 (2005), "Prequalified connections for special and intermediate steel moment frames for seismic applications", American Institute of Steel Construction, Chicago, IL.

[17] ANSI/AISC 360-05 (2010), "Specification for structural steel buildings", American Institute of Steel Construction, Chicago, IL.

[18] IS-8500 (1991), "Structural steel - micro alloyed (medium and high strength qualities) – specifications", Bureau of Indian Standards, New Delhi.

[19] Quimby TB (2011), "A Beginner's Guide to Structural Engineering", T. Bartlett Quimby, Inernet Version, 2011.

[20] Okazaki T, Liu D, Nakashima M, Engelhardt MD (2006), "Stability requirement for beams in seismic steel moment frames", Journal of Structural Engineering, Vol 132 (9), pages: 1334-1342.

NOTATION

F_y	=	The specified minimum yield stress of the material of the yielding element (beam/column).
E	=	Modulus of elasticity
P_a	=	Required axial strength of a column using ASD load combinations
P_u	=	Required axial strength using LRFD load combinations
P_y	=	Nominal axial yield strength of a member
C_a	=	Ratio of required strength to available strength
b_f	=	Width of flange
h	=	Clear distance between flanges less the fillet or corner radius for rolled shapes
t_w	=	Thickness of web
t_f	=	Thickness of flange
ϕ_c	=	Resistance factor for compression (0.9)
Ω_c	=	Safety factor for compression(1.67)
ϕ_b	=	Resistance factors for axial compression (0.9)
Ω_b	=	Factors of safety for axial compression (1.67)

POTENTIAL OF USING ROSA CENTIFOLIA TO REMOVE IRON AND MANGANESE IN GROUNDWATER TREATMENT

AeslinaBinti Abdul Kadir[1], Norzila Binti Othman[2], NurulAzimahBinti M.Azmi[3]

[1,2,3] Faculty of Civil and Environmental Engineering, UniversitiTun Hussein Onn Malaysia, Johor, Malaysia

*Corresponding E-mail : aeslina@uthm.edu.my[1]
norzila@uthm.edu.my[2]
nurulazimah86@yahoo.com[3]

ABSTRACT

Groundwater is source for water supply because of its good natural quality. However, groundwater may be exposed toward to contamination by various anthropogenic activities such as agricultural, domestic and industrial. Groundwater quality problem are typically associated with high hardness, high salinity and elevated concentration of iron, manganese, ammonium, fluoride and occasionally nitrate and arsenic. Therefore, groundwater should be treated to acceptable level before consumption. This study is carried out with the objectives to optimize the feasibility condition of contact time, biosorbent dosage and pH range in removing heavy metal by using Rosa Centifolia (R. Centifolia) and also to determine the water quality of groundwater sources. A dried Rosa Centifolia pretreated before being used as biosorbent. Experiment was done by varying contact time, biosorbent dosage and pH range to get the optimum value. The removal characteristic of Iron and Manganese by Rosa Centifolia was analyzed using Atomic Absorption Spectrophotometer (AAS). The optimum condition is achieved at 240 minutes, 0.05g/ml and pH 5 respectively. The optimum percentage removal of Iron and Manganese was found to be more than 70%. The finding indicated that Rosa Centifolia is a promising biosorbent in treating groundwater from RECESS UTHM wel.

Keywords: Groundwater, Rosa Centifolia, Fe, Mn, Biosorption

1.0 INTRODUCTION

Most groundwater contains some metal such as iron and manganese which naturally leaches from rocks and soils [1,2,3]. These metals that are found naturally, in soils, rocks, plants and most water supplies are essential for human health [4,5]. Excess amounts of metal in drinking water can cause coloured water, rusty-brown stains or black specs on fixtures and laundry [6,7]. Excess amounts of metal may also affect the taste of beverages and can build up deposits in pipes, heaters or pressure tanks [8]. The accumulation of metal contamination in water has become a concern due to the growing health risk is poses to the public [9,10]. Exposure to heavy metal contamination has been found to cause kidney damage, liver damage, and anemia in low doses [11,12,13].

Due to the severity of heavy metal contamination and potential adverse health impact on the public, efforts must be taken to purify water containing toxic metal ions. However, the conventional methods for removing heavy metals from groundwater such as precipitation and sludge separation, chemical oxidation or reduction, ion exchange, reverse osmosis, membrane separation, electrochemical treatment and evaporation are often ineffective and costly when applied to dilute and very dilute effluents [14,1,15]. Due to this problem, in this study, the biosorption method was used to remove the pollutants. Biosorption is one of the alternative method to remove heavy metal [19,20]. This method was chosen due to its economical and will reduce the production of sludge at the end of treatment process. Recently, researchers have proven to this alternative method by using many type of biomass waste that provides low cost maintenance but high removal efficiency of heavy metals [21,22]. Most studies prefer on choosing waste biomass that readily discard as waste in compare to costly biomass. R. Centifolia is one of the potential biomass that normally being discarded after extraction of essential oil that can be used as biosorbent to remove Fe and Mn in groundwater.

Commercially, R. Centifolia is used for its aesthetic value in perfumes manufacturing, decoration and healthy purpose. Normally, the number of rose consumptions is approximately 500 to 3000 petals roses in a year [23]. The R. Centifolia biomasses that were left after being used is a waste material with no commercial value or other usage. Hence, to reduce the waste of biomass, the R. Centifolia biomass was used as a biosorbent to solve this groundwater problem. The chemical composition of R. Centifolia waste biomass is described later on in this manuscript. In continuation of our investigation, the objective of this study was to determine the water quality of groundwater sources from RECESS, UTHM well. The effects of contact time, biosorbent dosage and pH range on Mn and Fe biosorption are described here in detail.

2.0   MATERIALS AND METHOD

This study consists of comprehensive batch experimental work. Figure 2.1 shows the R. Centifolia that was used in this experiment. Groundwater samples were collected at RECESS UTHM in Parit Raja for the study. The method that was used to collect the groundwater sample is the grab method. In the experimental work, five stages of experiments were conducted.

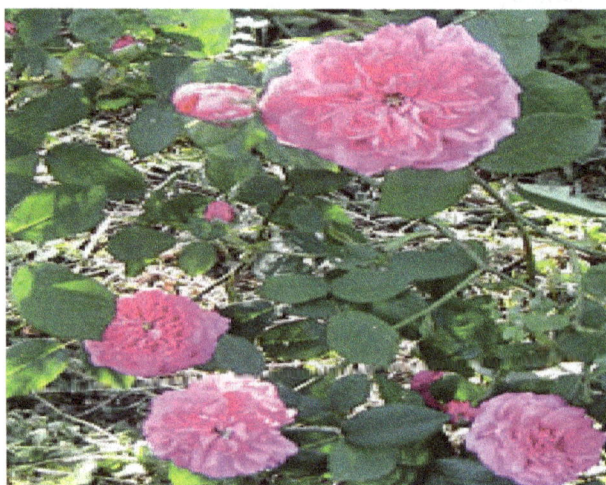

Figure 2.1: Rosa Centifolia

Figure 2.2 shows the step of experimental work. Firstly, the groundwater quality was tested to determine the culprit heavy metals [24]. At the second stage, the treatment to treat the groundwater pollution was chosen. In this study, biosorption treatment using R. Centifolia in Figure 2.2 was chosen as the biosorbent to treat the groundwater pollution. Third stage is the preparation of R. Centifolia as the biosorbent. The used R. Centifolia was collected and was sun dried in 7 days. Then it was thoroughly washed with running water and three times with distilled water to remove dirt. The washed R. Centifolia was oven dried at 60oC for 72 hours until constant weight. Then, it was cut into small pieces with homogenous known particle size of less than 0.255 mm [25] and turned into powder using mixer. In the fourth stage, optimum condition for the biosorbent to remove the heavy metals was determined by making the biosorbent dosage, pH and contact time as the variables in the batch reactor conducted in the laboratory. Batch study was carried out in this stage. Last stage (5th stage), the optimum condition determined in batch reactor was applied repeatedly to confirm and determine the effectiveness of the removal. The rose waste biomass samples were analyzed using X-Ray Fluorescence (XRF) for trace metals. Kinetic experiments (30–330 min) were performed to evaluate the effect of contact time [46]. The biomass (at 0.05 g/mL) was added into the conical flasks containing the 100mL of groundwater (100 mg/L) at pH 6. The flasks were agitated at 100rpm. In the dose (0.05–0.45 g/mL), size (less than 0.25mm), 100mL solution was taken in 250mL conical flask and were agitated at 100rpm for the initial contact time was maintained at the optimum value of 240min for Mn and Fe. In order to investigate the effect of pH in groundwater at various initial pH, it were prepared using 0.1MHCl or 0.1MNaOH. The optimum biomass for Mn and Fe (at 0.05 g/mL) was added into the conical flasks containing the 100mL of groundwater (100 mg/L). The flasks were agitated at 100rpm for 240 min. The metal concentrations were measured using atomic absorption spectrophotometer (AAS) (Perkin Elmer Analyst 300). The initial and final concentrations were determined using AAS. The amount of Fe and Mn taken up by rose waste biomass in each flask was calculated using following mass balance Equation 2.1 [26]:

$$\text{Percentage sorption} = \frac{Ci - Ce}{Ci} \times 100 \qquad \text{(Equation 2.1)}$$

Where:

Ci = initial and equilibrium metal concentrations (mg/l)
Ce = after treatment metal concentration (mg/l).

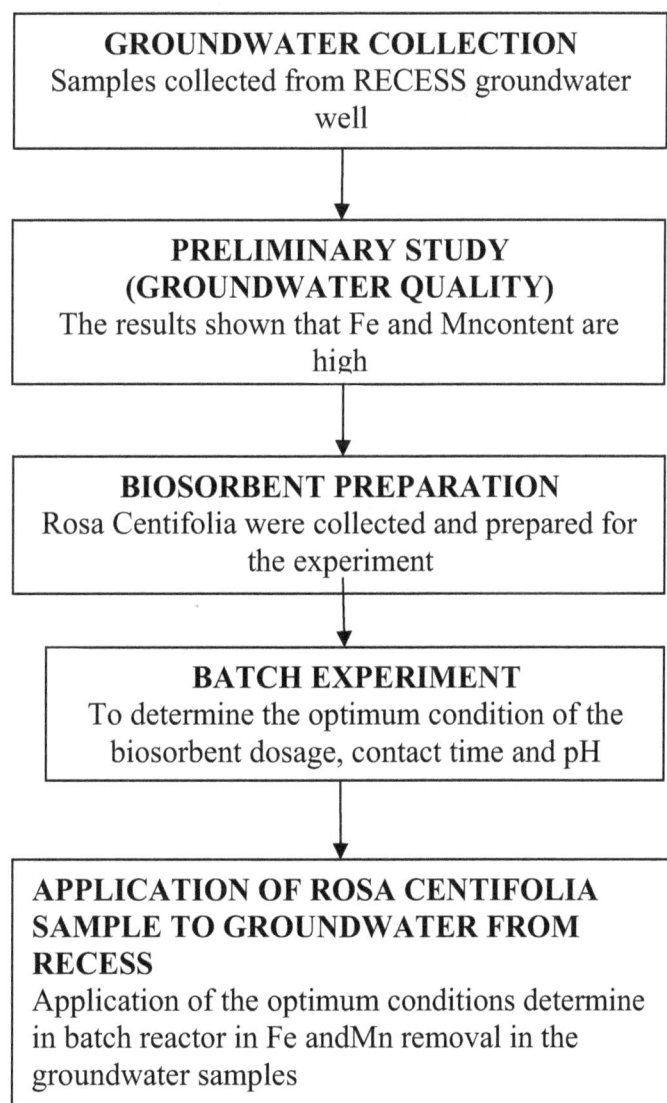

GROUNDWATER COLLECTION
Samples collected from RECESS groundwater well

↓

PRELIMINARY STUDY (GROUNDWATER QUALITY)
The results shown that Fe and Mncontent are high

↓

BIOSORBENT PREPARATION
Rosa Centifolia were collected and prepared for the experiment

↓

BATCH EXPERIMENT
To determine the optimum condition of the biosorbent dosage, contact time and pH

↓

APPLICATION OF ROSA CENTIFOLIA SAMPLE TO GROUNDWATER FROM RECESS
Application of the optimum conditions determine in batch reactor in Fe andMn removal in the groundwater samples

Figure 2.2: Experimental Work

3.0   RESULTS AND DISCUSSION

The chemical composition of R. Centifolia waste biomass is shown in Table 3.1. The obtained results clearly indicated that the rose waste biomass is cellulosic in nature and have replaceable hydrogen, alkali and alkaline metals. Thus, it can be regarded as potential Mn and Fe adsorbent. The effect of different experimental parameters on Mn and Fe biosorption is described in detail below.

Table 3.1: The elemental and proximate composition of Rosa centifolia waste biomass

Metal	Concentration (%)
CO_2	0.10
K_2O	62.90
CaO	9.72
P_2O_5	8.15
SO_3	6.52
S_iO_2	4.38
MgO	3.30
Fe_2O_3	1.76
Cl	1.01
Al_2O_3	0.80
ZrO_2	0.44
MnO	0.42
ZnO	0.28
Cr_2O_3	0.20
Rb_2O	0.13

3.1   FOURIER TRANSFORMS INFRARED (FTIR) STUDIES

FTIR analysis is an important analytical tool for determination of functional groups responsible for heavy metal removal by R. Centifolia. Figure 3.1 show the peak frequency of functional group that contain in R. Centifolia.

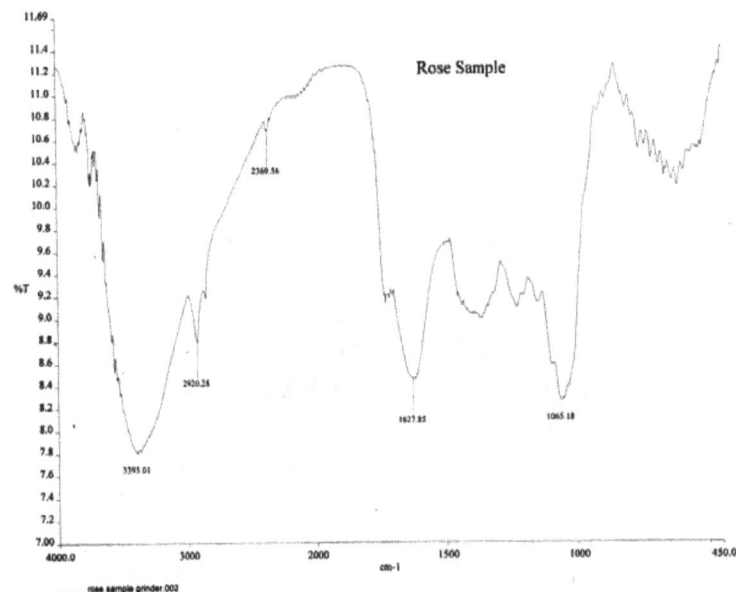

Figure 3.1: FTIR spectra of *R. Centifolia* waste biomass.

In order to determine the functional groups responsible for metal uptake, the FTIR was employed as an analytical [27]. Figure 3.1 shows the raw spectrum of waste biomass sample. The spectra indicate the presence of different functional groups which are responsible for metal sorption process for example carboxylic acids display a broad intense –OH stretching absorption from 3300 to 2500 cm−1, amines (3393cm−1), esters (1065cm−1), and alkenyl (1628cm−1). The peak observed at 2920.28cm-1 are due to vibrations of OH and CO groups belonging to carboxylic acids. Nasir (2007) had pointed out that peak at position 2920.28cm-1 shifts towards higher wave number, which indicates involvement of this group in removal metals. Ester groups in the biomass produce carboxylate, which can bind cations. Another study by Padmavathy (2003) further proved that the peak at 1065cm−1 also shift towards higher wave number, which indicates involvement of this group in removal metals. It is shown that, chemical functional group in the biomass tends to interact with the metals ion, and it caused stretching, contraction and bending of its chemical bonds. It is expected that functional groups in R. Centifolia have high ability to absorb Fe and Mn.

3.2   EFFECT OF CONTACT TIME

The optimum contact time was determined based on the percentage removal efficiencies. The condition which had the highest removal efficiencies of Fe and Mn had been chosen as the optimum condition. Figures 3.2 shows the percentage removal of Fe and Mn at different contact times.

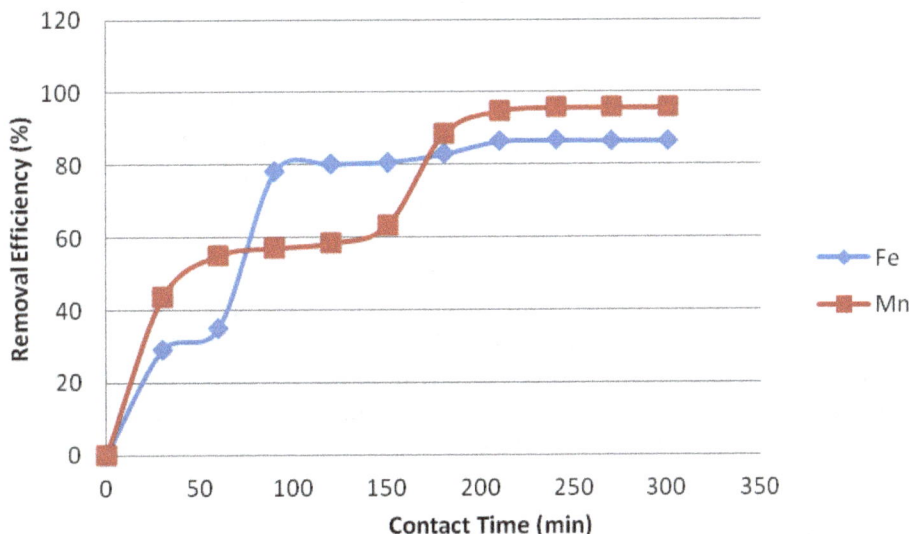

Figure 3.2: Effect of contact time to the removal efficiencies of Fe and Mn

From Figure 3.2 it was observed that the percentage removal of Fe and Mn ions by R. Centifolia increased with time and at 240min reached a constant value where no more removal increment were observed. At 240min the percentage removal of Fe is about 86.68% where for Mn the percentage removal is reached at 95.73%. At this point, the amount of metals being absorbed onto the biosorbent was in a state of dynamic equilibrium with the amount of absorb from the biosorbent [30]. The state of time required to attain this state of equilibrium was termed as the equilibrium time and the amount of metals absorbed at this equilibrium time reflected the maximum metals biosorption capacity of the biomass under the particular condition.

3.3   EFFECT OF BIOSORBENT DOSAGE

The optimum biosorbent dosage was also determined based on the percentage of the removal efficiencies. The condition which had the highest removal efficiencies of Fe and Mn had been chosen as the optimum condition. Figures 3.3 show the percentage removal of Fe and Mn at different biosorbent dosage.

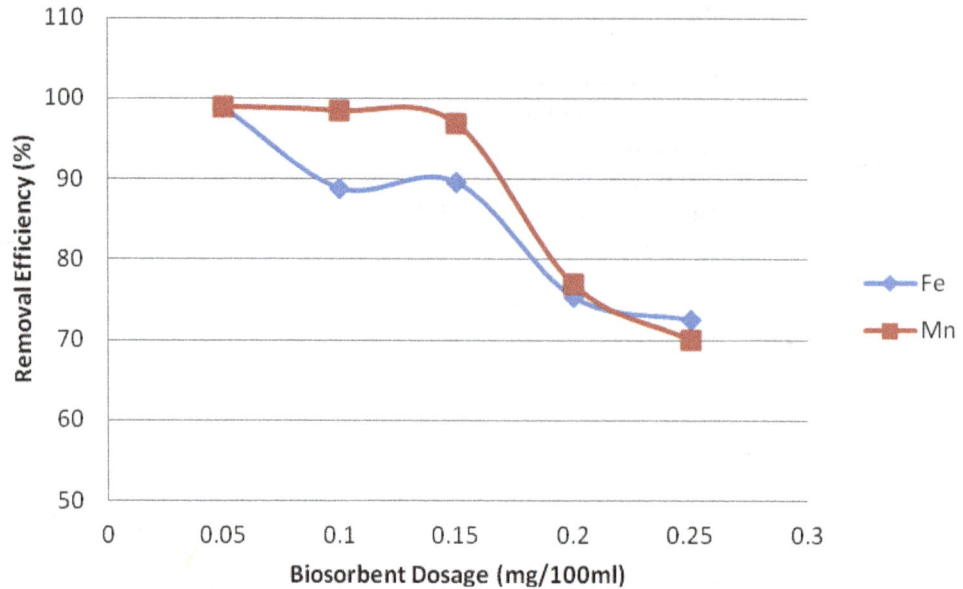

Figure 3.3: Effect of biosorbent dosage to the removal efficiencies of Fe and Mn

The results from Figure 3.3 shows that the maximum biosorption by immobilized R. Centifolia waste biomass occurred at 0.05 g/100 ml for both Fe and Mn. This shows that the maximum absorption occurs at minimum dose and hence the amount of metals bound to the absorbent and amount of free metals remained constant. After this dose, the uptake capacity (mg/100ml) of biosorbent was gradually decreased with increase in dose. Amount of biosorbent added to the solution determines the number of binding sites available for absorption. Similar results have been reported by Tariq (2010) and Bhatti (2007).

For effective metal sorption, biosorbent dose is a significant factor to be considered. It determines the sorbent–sorbate equilibrium of the system [25]. Moreover, as the biomass concentration rise, maximum biosorption capacity dropped, indicating poorer biomass utilization (lower efficiency). Results could be explained as a consequence of a partial aggregation, which occurred at high biomass concentration resulted to decrease of active sites.

3.4   EFFECT OF PH

Figures 3.4 shows the percentage removal of Fe and Mn at different pH value. From the graph it is shows that the higher the pH the higher the removal of Fe and Mn. Experiments concerning the effect of pH on sorption were carried out within pH range that was not influenced by metal precipitation (as metal hydroxide). The suitable pH ranges for two metal ions were slightly different like in the Figure 3.4 for Mn sorption were performed at the pH range of 1 to 5 and for Fe at pH of 1 to 6.

The biosorption of Mn and Fe on the R. Centifolia distillation sludge biomass was observed to be the function of solution pH. The pH is an important parameter in biosorption from aqueous solution because it influences equilibrium by affecting the speciation of the metal ions, solubility of the metal ions, concentration of counter ions on functional groups of the biomass and degree of ionization of the adsorbate during process (Say, 2003) and (Nadeem,2008). Percentage removal efficiency increased steadily with increasing pH. The most adequate sorption pH was 5 for Mn and 6 for Fe.

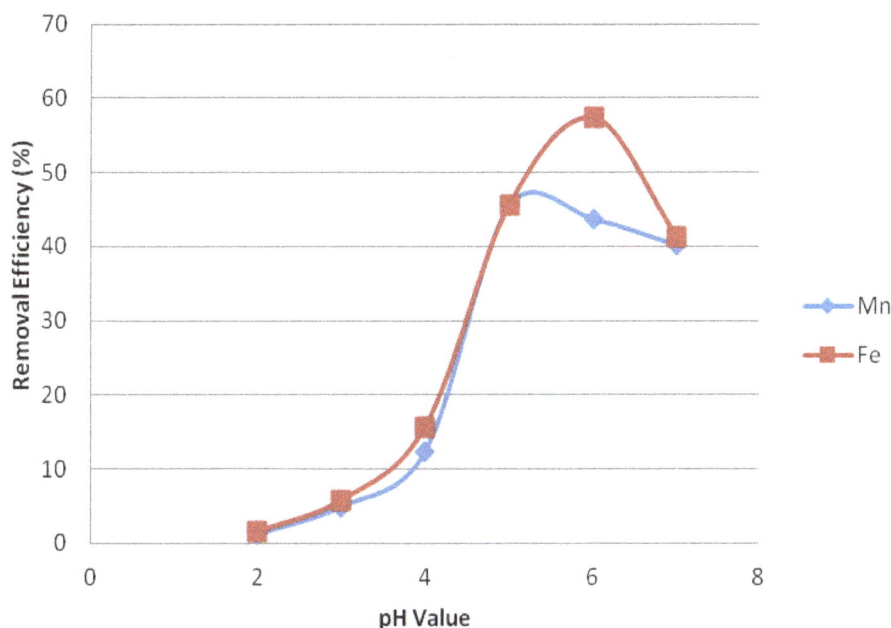

Figure 3.4: Effect of pH value to the removal efficiencies of Fe and Mn

The increase in biosorption with increase in pH can be explained by the fact that at low pH, the biosorbent surface became more positively charged thus reducing attraction between the biomass and metal ions. These bonded active sites thereafter became saturated and was inaccessible to other cations [35]. Moreover, as the pH is increased, the ligands such as carboxyl, sulfhydryl, phosphate others groups would be exposed, increasing the negative charge density on the biomass surface, resulting in greater attraction between metallic ions and ligands [35].

3.5　BATCH REACTOR PERFORMANCE

From batch experiment the optimum condition result is analysed. The optimum condition for contact time is 240min while for biosorbent dosage optimum condition is 0.05g/100ml for both Fe and Mn. For pH value, the optimum condition for percentage removal of both Fe and Mn is different which pH 5 is for Mn and pH 6 for Fe. For this study, pH 5 was selected to be used as the optimum condition. Table 3.2 showed the percentage removal of all parameter compare with the NDWQS using the optimum condition of contact time, biosorbent dosage and pH range.

Table 3.2: Percentage removal of all parameters from optimum condition

Parameters	Unit	Groundwater Sample			National Drinking Water Quality Standard (NDWQS)
		Before	After	Percentage Removal (%)	
pH	-	6.99	7.12	1.83	6.5-8.5
COD	mg/l	71	52.38	26.23	10
BOD	mg/l	52.99	32.28	39.08	1
TSS	mg/l	16.72	12.35	26.14	25
Nitrate	mg/l	5.1	26.4	-	10
Ammonia Nitrogen	mg/l	4.38	33	-	0.1
Fe	mg/l	2.37	0.009	99.55	0.3
Mn	mg/l	0.722	0.11	76.73	0.1

Table 3.2 shows that the percentage removals of Fe is higher than percentage removal of Mn with the value of 99.62% and 82.78% for Fe and Mn respectively. Both Fe and Mn are complying with NDWQS [36]. The percentage removal for pH is about 1.83% while for TSS is 26.14%. Both pH and TSS results are comply with NDWQS. Other parameter namely, COD, BOD, Nitrate and Ammonia Nitrogen are higher than acceptable level following NDWQS [37]. These results indicate that biosorbent only effective in removing of inorganic in groundwater sample.

3.5.1   IRON (FE) REMOVAL

Iron (Fe) being the fourth most abundant element and second most abundant metal in the earth's crust is a common constituent of groundwater [4,13]. According to the value in Figures 3.5, the influent data shows that Fe concentration in average 2.37mg/l. The high presence of iron in groundwater is generally attributed to the dissolution of iron bearing rocks and minerals, chiefly oxides (hematite, magnetite, limonite), sulphides, carbonates and silicates under anaerobic conditions in the presence of reducing agents like organic matter and hydrogen sulphide [39,40]. From Figure 3.5 the Fe concentration before the treatment is high compared to standard and it is out of the compliance range. After the treatment, the concentration of Fe is achieved the compliance range of the standard which is the Fe concentration is less than 1mg/l.

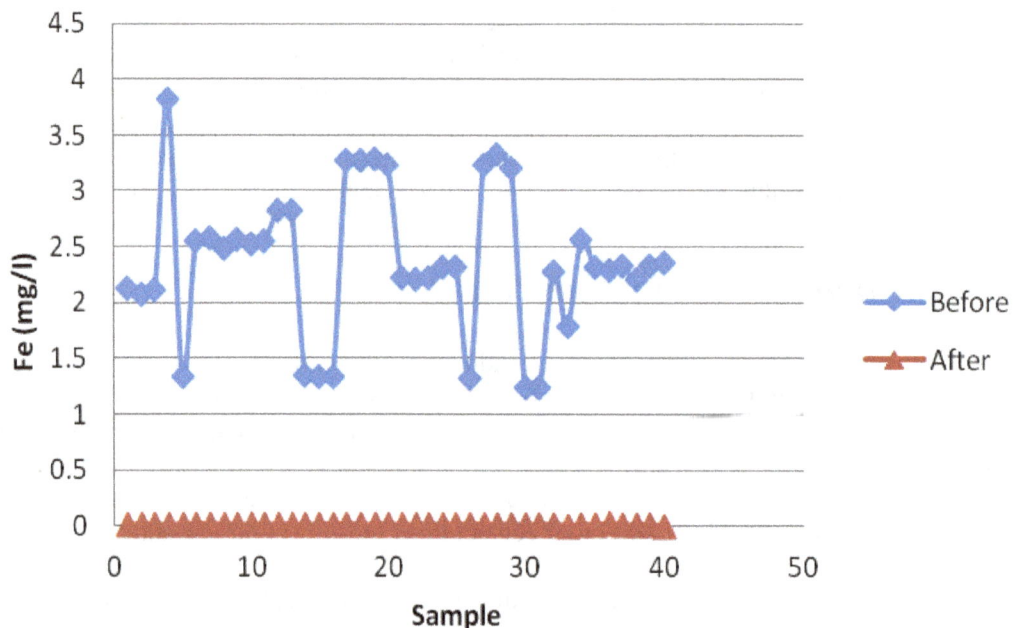

Figure 3.5: Average Iron (Fe) before and after treatment

Figure 3.6 shows that the percentage removal of Fe in this study was very high which is in average of 99.55%. According to Mubashir (2007) has obtained Pb (II) removals of 87.74% and Zn (II) with 73.8%. While Tariq (2010) has reported that Pb (II) removal is increasing to 95.67% by treating immobilized biomass with different chemical reagents (H2SO4, HCl and H3PO4). In this study, the results were in good agreement with previous researcher which is in the range of 70% to 88% ion metals removal [31].

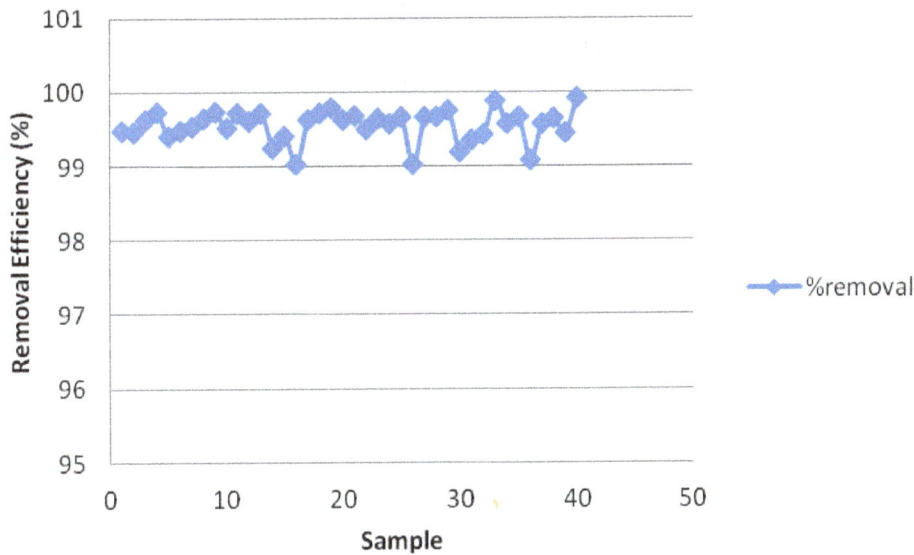

Figure 3.6: Graph of Iron (Fe) percentage removal

3.5.2   MANGANESE (MN) REMOVAL

Mn is a transition element that rarely exceeds 0.1 milligram per liter (mg/L) in natural water. In most of all the collected water samples there is a slight excessive level of Mn, hence there is an oily taste. According to the value in Figures 3.7, the influent data shows that average value Mn concentration is 0.73mg/l. The high presence of Mn causes adverse health effects in fact, essential to the human diet. However, water containing excessive amounts of Mn can stain clothes, discolor plumbing fixtures, and sometimes add a "rusty" taste and look to the water [42]. The Mn concentration before the treatment is high compare to standard and it is out of the compliance range. After the treatment, the concentration of Mn achieved the compliance with the standard which is the Mn concentration which is less than 0.1mg/l.

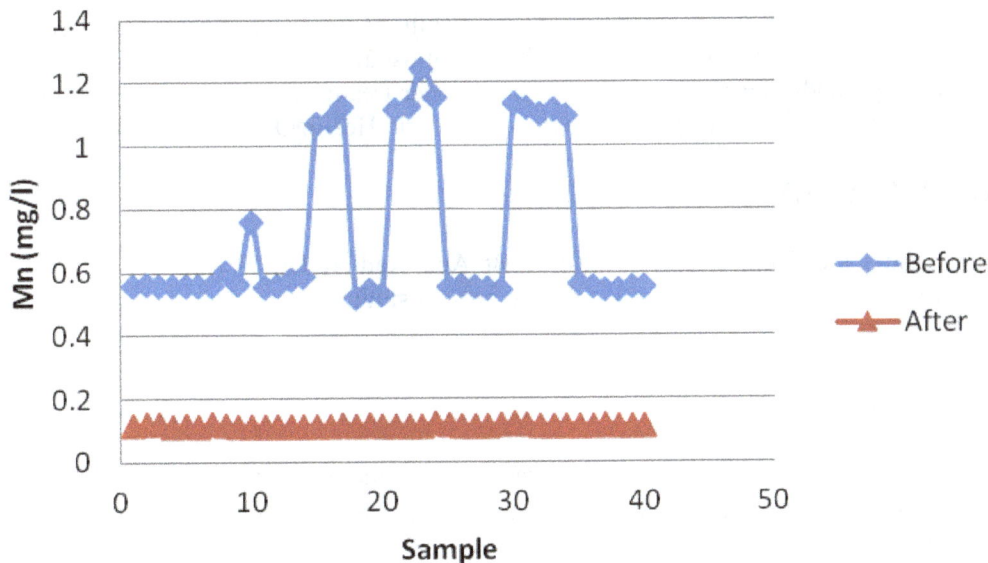

Figure 3.7: Average Manganese (Mn) before and after treatment

Figure 3.8 shows that the percentage of Mn removal in this study is in average of 76.73%. The maximum percentage removal of Mn is 89.93% while the minimum percentage removal is

64.14%. According to previous studies, (Abdul, 2009) have obtained Cu (II) removals of 55.79% and Sayed (2010) have reported Cu (II) 71.24% of removal. In this study, the results were in good agreement with those determined by other researcher which is in the range of 50% to75% of ion metals removal.

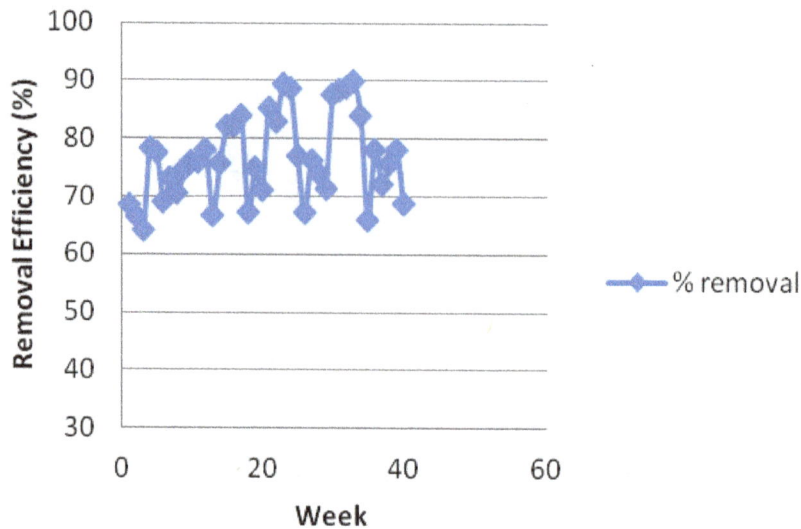

Figure 3.8: Graph of Manganese (Mn) percentage removal

4.0   CONCLUSIONS

This study concluded that R. Centifolia is an effective biosorbent in ferum and mangan removal. The optimum condition of heavy metal removal was found to be at 240min contact time, pH 5 and 0.05g/100ml dosage of R. Centifolia. The highest amount of Fe and Mn removal are 99.55% and 76.73% respectively. In this study parameter such as Fe, Mn, COD, BOD, pH and TSS show a reduce trend after the treatment, while other parameter such as Nitrate and Nitrogen Ammonia show unpromising result. These result show that R. Centifolia is effective to treat inorganic in compare to organic pollutant. The results of Fe and Mn removal in groundwater show that R. Centifolia is a promising biosorbent. Almost all parameter tested using groundwater show that the groundwater is safe to be consumed after treatment process using biosorbent as it follow acceptable level following limit provided by World Health Organization [41].

ACKNOWLEDGEMENT

Authors are thankful to Dr. Aeslina Bt Abd Kadir , Dr. Norzila Bt Othman and Pn Sabariah Musa from UTHM for their aid during present studies.

REFERENCES

[1]   Das Gupta A (1991), "Study on groundwater quality and monitoring in Asia and the Pacific". In : Groundwater Quality and Monitoring in Asia and the Pacific. Water Resources Series No 70, ESCAP, Bangkok, p 13-99.

[2]   EEA (1999), "Groundwater quality and quantity in Europe". June 1999, European Environmental Agency, Copenhagen, Denmark.

[3]   Sampat, P. (2000), "Groundwater shock". World Watch, January/February,10 22.

[4]   Fetter C.W. (2001), "Applied Hydrogeology" Fourth Edition, Prentice Hall, Inc.

[5] McCarty P.L, (1994), "An Overview of Anaerobic Transformation of Cholorinated Solvents". EPA Smposium on Intrinsic Bioremediation of Groundwater. Officeof research and Development, U.S. Environmental Protection Agency,Washington, D.C. p. 135-142

[6] Madhumita Das (2008), "Groundwater contamination due to manganese mining and its impact on health of mineworkers- a case study". Utkal University (India) Shreerup Goswami, Fakir Mohan University (India).

[7] Marcovecchio J.E, Botte S.E, Freije R.H (2007), "Heavy Metals, Major Metals, Trace Elements. In: Handbook of Water Analysis". L.M. Nollet, (Ed.). 2nd Edn. London: CRC Press, pp.275-311.

[8] Biswas, A.K. (1997), "Water Resources: Environmental Planning Management and Development". McGrawHill.

[9] Francesca, P. (2002), "Biosorption Of Binary Metal Systems Onto An Olive Pomace: Equilibrium Modelling". Department of Chemistry - University of Rome.

[10] Katko T.S, (1997), "Water! Evolution of Water and Sanitationin Finland from the mid-1800s to 2000". Finnish Water and Waste Water Association; Hameen Kirjapaino Ltd., Tampere.

[11] U.S. Environmental Protection Agency, (1992), "Lead Poisoning and Your Children, (800-B-92-0002)". Washington, D.C.: Office of Pollution Prevention and Toxics.

[12] Adeyemi O, Oloyede O.B, Oladiji, A.T (2007), "Physicochemical and Microbial Characteristics of Leachate-contaminated Groundwater". Asian. J. Biochem. 2(5): 343-348.

[13] Agency for toxic Substance and Disease Registry ASCE. (1990), "Toxicological profile for lead—draft. Atlanta: Evapotranspiration and irrigation water requirements". Manuals and reports on engineering practice. New York, NY, USA.

[14] Cynthia R. Evanko, David A. Dzombak. (1997), "Remediation of Metals Contaminated Soils and Groundwater prepared". P.E. Carnegie Mellon University Department of Civil and Environmental Engineering Pittsburgh.

[15] Adesola Babarinde N.A, (2007), "Isotherm and thermodynamic studies of the biosorption of copper (II) ions by Erythrodontium barteri". International Journal of Physical Sciences Vol. 2 (11), pp. 300-304.

[16] Adeyemi O, Oloyede O.B, Oladiji, A.T (2007), "Physicochemical and Microbial Characteristics of Leachate-contaminated Groundwater". Asian. J. Biochem. 2(5): 343-348.

[17] Delleur J.W. (1999), "The handbook of Groundwater Engineering". Indiana U.S.A. RC Press & Springer. Verlag Publisher.

[18] Mohammad Hassan Khani (2005), "Biosorption Of Uranium From Aqueous Solutions By Nonliving Biomass Of Marinealgae Cystoseira Indica". Atomic Energy Organization of Iran Jaber Ibn Hayan Research Laboratories Tehran, Iran.

[19] Siti Aishah A. W. (2009), "Biosorption Of Lead (II) From Aqueous Solution by Dried Water Hyacinth (Eichhornia Crassipes)". University Malaysia Pahang (Malaysia).

[20] Ozdemir G, Ozturk T, Ceyhan N, Isler R, Cosar T. (2003), "Heavy metal biosorption by biomass of Ochrobertrum anthropi". Proceeding exo-polysaccharide in activated sludge, Bioresour. Technol 90, pp: 71–74.

[21] Quek S.Y, Wase D.A.J, Forster C.F, (1998), "The use of sagowaste for the sorption of lead and copper". Water SA 24, pp: 251–256.

[22] Kratochvil D, Pimentel P.F, Volesky B, (2000), "Removal of trivalent chromium by seaweed biosorbent", Department of Chemical Engineering, McGill University,Montreal, Quebec, Canada,.

[23] Bhatti H.N, B. Mumtaz, M.A. Hanif, R. Nadeem, (2007), "Removal of Zn(II) ions from aqueous solution using Moringa oleifera lam. (Horseradish tree) biomass:, Process Biochem, 42, pp: 547–553.

[24] Narioka Hajime, (2003), "Changes in Level of Groundwater Table and its Conservation: The Case of the Ohba Magistrate Office, Setagaya-ku, Tokyo", J.Japan Soc. Hydrol. & Water Resour, 16 (6), pp: 631-639.

[25] Hanif M.A, Nadeem R, Zafar M.N, Akhtar K, Bhatti H.N. (2007), "Kinetic studies for Ni (II) biosorption from industrial wastewater by cassia fistula (Golden shower)". Biomass J. Hazard Mater 145, pp: 501-505.

[26] Javed MA, Bhatti HN, Hanif MA, Nadeem R (2007), "Kinetic and equilibrium modeling of Pb(II) and Co(II) sorption onto rose waste biomass". Sep Sci. Technol. In press.

[27] Silverstein R.M., F.X. Webster, (1998), "Spectrometric Identification of Organic Compounds". 6th ed., John Wiley & Sons,.

[28] Nasir MH, Nadeem R, Akhtar K, Hanif MA, Khaild AM (2007), "Efficacy of modified distillation sludge of rose (Rosa centifolia) petals for lead (II) and Zn (II) removal from aqueous solutions". J Hazard Mater; In press.

[29] Padmavathy V, Vasudevan P, Dhingra S.C, (2003), "Thermal and spectroscopic studies on sorption of nickel (II) ion on protonated baker's yeast". Chemosphere, 52, PP: 1807–1817.

[30] Haq Nawaz Bhattia, Rabia Khalida, Muhammad Asif Hanifa (2009), "Dynamic biosorption of Zn(II) and Cu(II) using pretreated Rosa gruss an teplitz (red rose) distillation sludge". Chemical Engineering Journal 148, pp: 434–443.

[31] Tariq Mahmood Ansari, (2010), "Immobilization of Rose Waste Biomass for Uptake of Pb(II) from Aqueous Solutions". Department of Environmental Sciences, GC University, Pakistan

[32] Bhatti H.N, B. Mumtaz, M.A. Hanif, R. Nadeem, (2007), "Removal of Zn(II) ions from aqueous solution using Moringa oleifera lam. (Horseradish tree) biomass". Process Biochem 42, pp: 547–553.

[33] Say R., N. Yilmaz, A. Denizli (2003), "Biosorption of cadmium, lead, mercury and arsenicions by the fungus Penicillium purpurogenum, Sep". Sci. Technol. 38, pp: 2039–2053.

[34] Nadeem R., Hanif M. A, Shaheen F, Perveen S, Zafar M. N, Iqbal T, (2008), "Physical and chemical modification of distillery sludge for Pb(II) biosorption". J. Hazard.Mater, 150, pp: 335–342.

[35] Yu J, M. Tong, X. Sun, B. Li, (2006), "Cystine-modified biomass for Cd(II) and Pb(II) biosorption", J. Hazard. Mater. 132, pp: 126–139.

[36] Nash, H. and McGall, G.J.H. (1994), "Groundwater Quality". 17th Special Report. Chapman & Hall.

[37] Hussein H, Ibrahim S.F, Kandeel K, Moawad H, (2004), "Biosorption of heavy metal from wastewater using pseudomonas sp". Eur. J. Biotechnol, 7, pp: 1–7.

[38] Hem, J.D. (1989), "Study and Interpretation of the Chemical Characteristics of Natural Water". 3rd ed, USGS, Water Supply Paper 2254.

[39] Nelder J.A, R.Mead, (1965), "A simplexmethod for function minimization". Comput. J. Vol. 7, pp: 308–315.

[40] Nomanbhay S.F. , K. E. J. (2005), "Palanisamy", Biotechnol. 8, 44,

[41] WHO (1996), "Guidline for the Promotion of Human Rights of persons with mental disorders". Geneva, World Health Organization.

[42] Adepoju-Bello A.A, Ojomolade O.O, Ayoola G.A, Coker AAB (2009), "Quantitative analysis of some toxic metals in domestic water obtained from Lagos metropolis". The Nig. J. Pharm., 42 (1): 57-60.

[43] Abdur Rauf Iftikhar (2009), "Kinetic and thermodynamic aspects of Cu(II) and Cr(III) removal from aqueous solutions using rose waste biomass". Industrial Biotechnology Laboratory, Department of Chemistry, University of Agriculture, Faisalabad, Pakistan.

[44] G.O El-Sayed (2010), "Biosorption Of Ni (II) And Cd (II) Ions From Aqueous Solutions Onto Rice Straw". Chemistry Department, Faculty of Science, Benha University, Benha, Egypt.

[45] Mubashir Hussain Nasir (2010), "Efficacy of modified distillation sludge of rose (Rosa centifolia) petalsfor lead(II) and zinc(II) removal from aqueous solutions". Department of Chemistry, University of Agriculture, Faisalabad (38040), Pakistan.

[46] Keskinkan O. (2003), "Heavy metal adsorption properties of a submerged aquatic plant (Ceratophyllum demersum)", Department of Environmental Engineering, Faculty of Engineering, Cukurova University, Turkey

13

COMPRESSIVE STRENGTH AND STATIC MODULUS OF ELASTICITY OF PERIWINKLE SHELL ASH BLENDED CEMENT CONCRETE

Akaninyene A. Umoh[1], K. O. Olusola[2]

[1]Building Department, University of Uyo,Uyo, Nigeria
[2] Building Department, Obafemi Awolowo University, Ile-Ife, Nigeria

*Corresponding E-mail : aumoh@ymail.com

ABSTRACT

The study examined the effect of periwinkle shell ash as supplementary cementitious material on the compressive strength and static modulus of elasticity of concrete with a view to comparing it's established relation with an existing model. The shells were calcined at a temperature of 800°C. Specimens were prepared from a mix of designed strength 25N/mm^2. The replacement of cement with periwinkle shell ash (PSA) was from 0 to 40% by volume. A total of 90 cubical and cylindrical specimens each were cast and tested for 7 to 180 hydration periods. The results revealed that the PSA met the minimum chemical and physical requirements for class C Pozzolans. The compressive strength and static modulus of elasticity were observed to increase with increased in curing age but decreased with increasing PSA content. The design strength was attained with 10%PSA at standard age of 28 days. In all the curing ages, 0% PSA content recorded higher value than the blended cement concrete. the statistical analysis indicated that the percentage PSA replacement and the curing age have significant effect on the properties of the concrete at 95% confidence level. The relation between compressive strength and static modulus of elasticity fitted into existing model for normal-weight concrete.

Keywords: *Blended cement; Compressive strength; Periwinkle shell ash; Static modulus of elasticity*

1.0   INTRODUCTION

Modern pozzolanic cements are mix of natural or artificial Pozzolans and Portland cement. in addition to under water use, the pozzolan's high alkalinity makes it especially resistant to common forms of corrosion from sulphates [1]. Once fully hardened, the Portland cement blended with Pozzolans may be stronger than Portland cement due to its lower porosity, which also makes it more resistance to water absorption and spalling [2]. The uses of Pozzolans in concrete have been reported to mitigate the effect of sulphates and alkali – silica reaction, especially deleterious in concrete structures, by the development of a faster pozzolanic reaction [2].

The commonly used Pozzolans have been fly ash, silica fume, metakaolin, and blast furnace slag. Recently, efforts have also been made in the use of agricultural wastes as Pozzolans. Some of the Pozzolans of agricultural origin include sawdust ash [3], rice husk ash [4], corn cob ash [5-6], millet husk ash [7], palm oil fuel ash [8] and periwinkle shell ash [9-10].

The processing and utilization of periwinkle shell, which is cheap and abundantly available in most riverine communities of Niger delta region of Nigeria, in construction; apart from improving certain properties of concrete will as well protect the environment from pollution, and thereby contributing to resource conservation and environmental sustainability. Periwinkle Shell Ash is obtained by burning periwinkle shell which is the by-product of Periwinkle.

Periwinkle is described as any small greenish marine snail from the class of gastropod, the largest of the seven classes in the phylum mollusc [11]. They are herbivorous and found on rocks, stones or pilings between high and low tide marks; on mud-flats as well as on prop roots of mangrove trees and in fresh and salt water.

The common periwinkle (*Littorina littorea*) is one of the most abundant marine gastropods in the North Atlantic, but *Tympanotonus fuscatus* is commonly found in the estuaries and mangrove swamp forest of the Niger Delta of the South – South region of Nigeria [12]. A study [13] indicated that there are about 40.3 tonnes of periwinkle per year being harvested from 35 mangrove communities of Delta and Rivers states of Nigeria. A survey, by the researchers, of some riverside communities of Itu, Oron, Issiet, Okobo, Ikot Offiong, and Uta-ewea in Akwa Ibom state showed abundance of periwinkle in these communities. Massive periwinkle harvesting is also reported from some communities in Bayelsa, Cross River and Edo states of Nigeria [14-15]. When the periwinkle is big enough, the edible part is removed after boiling in water, and the shell dumped as waste. The continuous dumping of the shells [16] has resulted in great heaps constituting menace especially in villages in Rivers and Akwa Ibom states of Nigeria. Therefore this work examined the possibility of processing and utilization of the shell ash with a view to establishing its performance on the compressive strength and static modulus of elasticity when used as supplementary cementitious material in concrete production.

2.0   LITERATURE REVIEW

The compressive strength of concrete is considered one of the most important properties in the hardened state. Neville [17] opined that for the purpose of structural design, the compressive strength is the criterion of quality. However, when the deformations of the different structural elements of a structure have to be calculated, the determination of elastic properties of concrete is also very important [18]. It is also stated [19] that the modulus of elasticity of concrete is one of the most important mechanical properties of concrete since it impacts the serviceability and the structural performance of reinforced concrete structures. The elastic characteristics of a material are a measure of its stiffness. Knowledge of the modulus of elasticity is essential in the determination of deformation, deflection or stresses under short-term and long-term loading [20]. Pozzolanic materials which are either by-products from industry or wastes from agricultural activities had gained wider acceptability in improving properties of concrete and other cement products. Pozzolans have been used to improve the compressive strength performance of concrete as it can continue to react for many years and thereby making the concrete harder and more durable during its service life. Pozzolans as partial replacement of cement in concrete have been reported to have slow strength gain especially during the early ages, a situation that makes it usage not feasible where early strength is of paramount importance. The slow contribution of pozzolanic materials to strength development of concrete have been attributed to the pozzolanic reaction at room temperature which is slow; and therefore a long curing period is needed to observe its positive effects [21]. Available studies on the use of periwinkle shell ash as pozzolanic material and its effects on concrete compressive strength only considered curing age up to 28 days, and noted that the compressive strength of the concrete increases with curing age but decreases as the PSA content increases, and that PSA replacement of OPC up to 40% was found adequate for the production of medium strengths concrete [9-10].

Also the uses of Pozzolans as replacement of cement have also been reported to affect the static modulus of elasticity of concrete [22-24]. For instance the use of rice husk ash in concrete as reported [22] contributed to higher value in static modulus of elasticity when compared to the reference concrete. on the other hand, Wainwright and Tolloczko [23] reported that the replacement of Portland cement by slag in concrete seems to decrease the modulus of elasticity for a compressive strength below $55N/mm^2$ and slightly increase it, by about 10%, for compressive strength greater than $60N/mm^2$. Modulus of elasticity is reported to be low at early ages and high at later ages for fly ash-blended cement concrete [24]. Therefore, the study examined the performance of PSA on the compressive strength and static modulus of elasticity when used as supplementary cementitious material in concrete.

3.0   METHODOLOGY

3.1   MATERIALS

Ordinary Portland cement (OPC) produced by Calcemco (Nig.) Ltd. to the specification of NIS 444 - 1 [25] and branded as 'UNICEM' was used. The periwinkle shells for the production of periwinkle shell ash (PSA) were collected from one of the dumpsites in Otto market in Ikot Ekpene of Nigeria. The shells were washed off unwanted dirt and dried in an open space before calcined in a gas furnace and stopped as soon as the temperature reaches 800^0C. The ash was ground and sieved with 45μm size. The specific gravities for cement and PSA were 3.13 and 2.13 respectively. Chemical and physical properties of the cementitious materials are shown in Table 1. The Strength activity index with Portland cement, which expresses the compressive strength of mortar cube when Portland cement is blended with 20% pozzolanic material by weight, was calculated as the percentage ratio (A/B) ×100 where A = average compressive strength of test mixture cubes, N/mm^2, and B = average compressive strength of control mix cubes, N/mm^2.

Fine aggregate used were river-bed sand passing 4.75mm sieve and falls within zone 2 with a fineness modulus of 3.28; while the coarse aggregate were crushed granite of maximum size 20mm with specific gravity of 2.65. The sieve analyses conducted on the aggregates are presented in Tables 2 and 3.

Table 1: chemical and physical properties of PSA

Chemical composition												
Elemental Oxide (%)	SiO_2	Al_2O_3	Fe_2O_3	CaO	MgO	SO_3	K_2O	Na_2O	P_2O	Mn_2O_3	TiO_2	LOI
	33.84	10.20	6.02	40.84	0.48	0.26	0.14	0.24	0.01	0.00	0.03	7.60
Physical properties												
% retained on 45μm sieve	Strength activity index with Portland cement (% of control)		Water used (% of control)		Soundness (mm)		Moisture content (%)		Specific gravity			
	7 days	14 days										
21.00	78.17	79.12	104		1.00		1.50		2.13			

Table 2: Sieve analysis of the sand

Sieve size (mm)	Weight of material retained (g)		Cumulative percentage of material retained	Percentage passing
	(g)	(%)		
4.75	0.0	0.0	0.0	100.0
2.36	19.0	3.82	3.82	96.18
1.18	80.5	16.16	19.98	80.02
0.600	185.0	37.15	57.13	42.87
0.300	173.0	34.74	91.87	8.13
0.150	35.0	7.03	98.90	1.10
Pan	5.5	1.10	100	0.00
Total	498.0	100	-	-

Fineness Modulus = (100 + 96.18 + 80.02 + 42.87 + 8.13 + 1.10)/100 = 3.28

Table 3: Sieve analysis of coarse aggregate

Sieve size (mm)	Weight of material retained (g)		Cumulative percentage of material retained	Percentage passing
	(g)	(%)		
20	0.0	0.0	0.0	100.0
14	240.0	16.00	16.00	84.0
9.5	496.0	33.09	49.09	50.91
4.75	670.6	44.74	93.83	6.17
2.36	80	5.34	99.17	0.83
Pan	12.4	0.83	100	0.00
Total	1499.0	100.0	-	-

2.2   PROPORTION AND MIXING OF CONSTITUENTS

The mix proportion involved the British mix-design method (Department of Environment- DOE) approach for 28-day characteristics strength of $25N/mm^2$ for Normal-weight concrete (i.e. control: 0% PSA and 100% OPC). A water/cement ratio requirement based on the strength of the Mix design for the requisite workability (slump: 10-30mm) was adhered. Partial replacement of OPC by PSA of various percentages of 0, 10, 20, 30 and 40 by volume as was dictated by their differences in specific gravities was adopted. The ingredients, that is, cement, PSA, aggregates and water, were manually mixed. The cement and PSA blended was spread on already measured sand, and the three ingredients mixed thoroughly before the coarse aggregate and water were added. Slump and compacting factor tests were carried out to determine the workability of each mix. The mix proportions of the mixes are presented in Table 4

Table 4: Mix proportions (m^3) of PSA blended cement concrete

PSA content (%)	Cementitious Binder (kg)		Water (kg)	Fine aggregate (kg)	Coarse aggregate (kg)	
	OPC	PSA			5-10mm	10-20mm
0	19.18	-	11.05	42.92	27.46	55.74
10	17.26	1.30	10.97	42.92	27.46	55.74
20	15.34	2.60	10.80	42.92	27.46	55.74
30	13.43	3.90	10.70	42.92	27.46	55.74
40	11.51	5.20	10.63	42.92	27.46	55.74

3.3   SPECIMENS PREPARATION

Two types of specimens were prepared: 150mm cubes and 150mm by 300mm cylinders. The casting was done as specified by BS EN 12390 part 3 [26]. As soon as the specimens were cast, they were stored in a place free from vibration and not exposed to direct sunlight or other sources of heat and covered with wet wooden bags. The specimens were de-moulded after 24 hours, placed in water curing tanks kept at temperature of $29\pm1°C$ until the testing age

3.4   TESTING

The concrete specimens were tested for cube compressive strength and static modulus of elasticity [27-28]. The cylindrical specimens were placed inside compressometer fixed with dial gauges, and the whole assembly mounted on a compression testing machine. The tests were done for six curing levels of 7, 14, 28, 90, 120 and 180 days for each of the five levels of PSA replacement of cement of 0, 10, 20, 30 and 40% respectively. At least three specimens were tested at each age for compressive strength and static modulus of elasticity to compute the average. A total of 90 cubes and cylinders each were used to determine the effect of PSA on the compressive strength and static modulus of elasticity for the various curing age. The tests were carried out using compression testing machine of 2000KN capacity.

4.0   RESULTS AND DISCUSSIONS

4.1   EFFECT OF PSA ON WORKABILITY

The results of the slump and compacting factor values are shown in Table 5. To attain the same workability level of 10-30mm in the mixes containing PSA with that of conventional concrete (i.e. control), higher water content was required. This is reflected in the gradual increased in the water cementitious material ratio with a corresponding increased in the amount of water over control as the PSA percentage content increases. This higher water requirement in mixes containing PSA could be attributed to the high fineness of PSA which meant a greater specific surface to be wetted and lubricated. This agreed with the earlier finding of the effect of rice husk ash in concrete [4]. The values of the slump range between 25 and 29mm which is within the standard required values of 10 – 30mm based on the mix design for concrete of low workability, while that of compacting factor value is between 0.83 and 0.87 and satisfied the range of 0.85 to 0.90 of compacting factor value for the same concrete [17].

Table 5: Slump and Compacting factor values for PSA blended cement concrete

PSA content (%)	Slump (mm)	Compacting factor	Actual water/cementitious material ratio	Amount of water over control (%)
0	29	0.86	0.58	-
10	28	0.87	0.59	101.72
20	28	0.85	0.60	103.45
30	26	0.84	0.62	106.29
40	25	0.83	0.64	110.34

4.2   COMPRESSIVE STRENGTH

The compressive strength development at various ages is given in Table 6. The compressive strength generally increased with curing age and decreased with increased content of periwinkle shell ash. The results at 7 days show that in all the replacement levels the percentage attainment of the design strength range between 77.63% and 65.66% with 0%PSA content (i.e. control) having 77.63% and 40%PSA content having the least value of 65.66%. These values satisfied the requirement of normal-weight concrete strength development which is stipulated to be between 50-66% [29-30].

At 14 days, the compressive strength of the control mix is $27.11N/mm^2$, representing 108.44% of the design strength, closely followed by 10%PSA which had 85.33% of the design strength; while 10, 20 and 30%PSA replacement had compressive strength of 18.04, 17.01 and $16.30N/mm^2$ which is also 72.18, 68.04 and 65.19% of the design strength respectively. The strength development at 14 days satisfied the 60-75% of the design strength as stipulated [29].

The compressive strength of 0% and 10%PSA content at 28 days hydration period were $28N/mm^2$ and $25.56N/mm^2$ respectively which met the desired design strength of $25N/mm^2$, while that of 20, 30 and 40%PSA content were 24.15, 20.71 and $15.91N/mm^2$ respectively. These are comparable with the values obtained by other researchers [9-10]. The strength development for control mix (i.e. 0%PSA) is faster up to 28 days hydration period whereas mixes containing PSA is slower. This portray the fact that the pozzolanic reaction depends on the released of calcium hydroxide from cement hydration.

The results at 90 days indicated that in all the mixes there is continuous increase in the strength, showing that there is both hydration and pozzolanic reactions particularly with 10%PSA having a higher rate of development than the control. At 120 days, 10%PSA recorded compressive strength of 28.53N/mm^2 representing an increase of 6.89% of the strength at 90 days, while the control mix recorded strength of 29.92N/mm^2 which represents an increase of 3.26% of the strength of 90 days. Other mixes had little or no increase in the design strength beyond 90 days. The 20%PSA had strength of 24.89N/mm^2 representing 99.56% (approximately 100%) of the design strength. It means that where later age strength is required at 120 days hydration period, 20% replacement of cement with PSA is adequate.

A further increased in the rate of strength development was observed with 10%PSA at 180 days as it attained strength of 29.04N/mm^2 which is not significantly different from the control which had strength of 30.15N/mm^2. The continuous increased in the 10%PSA can be attributed to the fact that the quantity of calcium hydroxide liberated from cement hydration is adequate to be consumed by the pozzolanic reaction.

Table 6: Compressive strength of PSA blended cement concrete specimens at all curing ages

Curing Age (Days)	PSA (%)	Compressive strength (N/mm^2)				Attainment of Design strength (%)
		Sample 1	Sample 2	Sample 3	Mean	
7	0	19.56	19.38	19.38	19.41	77.63
	10	18.22	18.67	18.67	18.52	74.07
	20	17.56	17.78	18.22	17.85	71.41
	30	17.33	17.69	17.51	17.51	70.04
	40	16.44	16.27	16.53	16.41	65.66
14	0	27.11	26.67	27.56	27.11	108.44
	10	21.33	20.80	21.87	21.33	85.33
	20	18.22	17.78	18.13	18.04	72.18
	30	17.33	16.71	16.98	17.01	68.04
	40	16.36	16.44	16.09	16.30	65.19
28	0	28.00	27.78	28.22	28.00	112.00
	10	25.78	25.33	25.56	25.56	102.22
	20	24.00	24.44	24.00	24.15	96.59
	30	20.89	20.44	20.80	20.71	82.84
	40	15.56	16.18	16.00	15.91	63.64
90	0	28.89	29.11	29.33	29.11	116.44
	10	26.67	26.76	27.02	26.81	107.24
	20	24.44	24.89	24.89	24.74	98.96
	30	21.33	20.89	21.56	21.24	84.98
	40	17.33	17.78	17.78	17.63	70.52
120	0	29.78	30.22	29.78	29.92	119.70
	10	28.44	28.44	28.71	28.53	114.13
	20	24.89	24.89	24.89	24.89	99.56
	30	20.00	21.33	19.56	20.30	81.19
	40	16.89	17.78	17.33	17.33	69.33
180	0	30.22	30.00	30.22	30.15	120.59
	10	29.33	28.89	28.89	29.04	116.15
	20	24.00	23.78	23.56	23.78	95.11
	30	21.78	21.33	22.22	21.78	87.12
	40	20.89	20.67	20.44	20.67	82.67

4.3   STATIC MODULUS OF ELASTICITY

The results of the Static Modulus of Elasticity are presented in Table 7. At 7 days, the values are 24359 N/mm^2, 24115 N/mm^2, 23872 N/mm^2, 23209 N/mm^2 and 20042N/mm^2 for 0, 10, 20, 30 and 40% PSA replacement of cement respectively. The elasticity was observed to increase to 27032 N/mm^2, 25312 N/mm^2, 23842 N/mm^2, 23250 N/mm^2 and 21923N/mm^2 for 0, 10, 20, 30 and 40%PSA content respectively at 14 days curing period. At 28 days, the elasticity of concrete containing different percentages of PSA was noted to increase at a faster rate than the control. For instance, the values increased at 5.13%, 5.39%, 10.31%, 9.39% and 4.60% for 0, 10, 20, 30 and 40%PSA content respectively. All the mixes at 28 days met the requirement of 18,000N/mm^2 to 30,000N/mm^2 stipulated by BS 8110 part 2 [30] and that of 14,000N/mm^2 to 42,000N/mm^2 [31].

There was an elasticity improvement with 0% and 10%PSA blended cement concrete at 90 days, but a reduction was recorded with 20%, 30% and 40%PSA content. This an indication that there is continuous hydration and pozzolanic reactions with the blended cement concrete of 0% and 10%PSA content as evidence in higher rate of percentage increased.

The elasticity values at 120 days range between 30,210N/mm^2 and 20,479N/mm^2 for 0 - 40%PSA substitution respectively. At 180 days, there was no significant difference between the values recorded for 0% and 10% PSA content. This was closely followed by 20% PSA with a value of 28,208N/mm^2. Generally, the results revealed that the value of the static modulus of elasticity of the control (i.e. 0%PSA) is greater than those of the blended cement concrete in all the curing ages; and that the increased in the value with curing age, particularly with 0 – 20%PSA content, indicated the fact that there is a continuous hydration and pozzolanic reactions. The reduction in elasticity with increase in PSA content could be attributed to higher carbon content (expressed as loss on ignition) in the periwinkle shell ash and low quantity of cement in the mixes as a result of its replacement. This finding agreed with the use of slag in concrete [23] which recorded decreased in modulus of elasticity for compressive strength below about 55N/mm^2 and a slight increase (by about 10%) for compressive strength greater than about 60N/mm^2.

Table 7: Static modulus of elasticity of PSA blended cement concrete at different curing ages

PSA content (%)	Static Modulus of Elasticity (N/mm^2)					
	7 days	14 days	28 days	90 days	120 days	180 days
0	24359	27032	28419	29028	30210	31302
10	24115	25312	26676	27823	29846	30937
20	23872	23842	26299	25655	27215	28208
30	23209	23250	25434	24952	22483	22138
40	20042	21923	22932	21750	20479	18934

The statistical analysis, using analysis of variance (ANOVA), on the effect of PSA content and Curing age on the Static modulus of elasticity indicated that the independent factors (i.e. PSA content and curing age), when considered individually and collectively had significant effects on the Static modulus of elasticity of the concrete (Table 8). The coefficient of determination (adjusted R-Square value) is 0.969 (96.9%). This implies a strong statistical association among the variables. The independent variables were estimated to account for 96.9% of the variance in the Static modulus of elasticity of the concrete. The coefficient of correlation was obtained as R = 0.984. This shows that a very strong linear relationship exist between the two sets of variable being considered.

Table 8: Results of Anova for static modulus of elasticity

Source	Type III Sum of Squares	df	Mean Square	F	Sig.	Partial Eta Squared
Corrected Model	9.396E8	29	3.240E7	96.250	.000	.979
Intercept	5.746E10	1	5.746E10	170705.050	.000	1.000
PSA	6.420E8	4	1.605E8	476.823	.000	.970
CURAGE	1.216E8	5	2.431E7	72.224	.000	.858
PSA * CURAGE	1.760E8	20	8799457.619	26.141	.000	.897
Error	2.020E7	60	336610.722	-	-	-
Total	5.842E10	90	-	-	-	-
Corrected Total	9.598E8	89	-	-	-	-

4.4 RELATIONSHIP BETWEEN COMPRESSIVE STRENGTH AND STATIC MODULUS OF ELASTICITY

Table 9 shows the Static modulus of elasticity and cube Compressive strength for various PSA replacement levels with cement for all the hydration periods. It indicated that the modulus of elasticity increases with an increase in the compressive strength; and also increased at a higher rate than the compressive strength at later ages (i.e. 120 days and above). This agreed with earlier findings [31-32] which they attributed to the beneficial effect of improvement in the density of the interfacial transition zone, as a result of slow chemical interaction between the alkaline cement paste and aggregate, which is more pronounced for the stress – strain relationship than for the compressive strength of concrete.

Regression analyses performed to establish empirical relationship between the cube compressive strength and Static modulus of elasticity of concrete incorporating different percentages of periwinkle shell ash replacing cement was based on the expression $E = 9.1F_{cu}^{0.33}$ proposed by BS 8110 part 2 [30]. The regression equations were as follow:

0%PSA: $E_s = 9.464F_{cu}^{0.33}$; ($R^2 = 0.754$) ----- (1)

10%PSA: $E_s = 9.432F_{cu}^{0.33}$; ($R^2 = 0.772$) --- (2)

20%PSA: $E_s = 9.215F_{cu}^{0.33}$; ($R^2 = 0.717$) --- (3)

30%PSA: $E_s = 8.720F_{cu}^{0.33}$; ($R^2 = 0.716$) --- (4)

40%PSA: $E_s = 8.099F_{cu}^{0.33}$; ($R^2 = 0.86$) ---- (5)

The regression equation in each percentage PSA content are $9.464F_{cu}^{0.33}$, $9.432F_{cu}^{0.33}$, $9.215F_{cu}^{0.33}$, $8.720F_{cu}^{0.33}$ and $8.099F_{cu}^{0.33}$ for 0, 10, 20, 30 and 40%PSA content respectively. It indicated that the elasticity decreases with increase in the percentage of periwinkle shell ash. The coefficient of association, r has a value between 0.868 and 0.927, indicating a strong linear relationship between the two variables. The relation between the static modulus of elasticity and cube compressive strength at all curing ages gave a regression equation of $E_s = 9.050F_{cu}^{0.33}$ with $R^2 = 0.651$ (equation 6) and a corresponding r value of 0.807 showing a strong linear relationship. These equations(1 – 6) do not significantly differ from the equation for normal- weight concrete, therefore, for PSA blended cement concrete, a static modulus of elasticity at age up to 180 days can be predicted by the model given by BS 8110 part 2 [30] for normal- weight concrete.

Table 9: Static modulus of elasticity, Compressive strength and Cube root of compressive strength of PSA blended cement concrete for different curing ages

PSA content (%)	Curing age (Days)	Static modulus of elasticity (Gpa)	Compressive Strength (N/mm^2)
0	7	24.359	19.41
10		24.115	18.52
20		23.872	17.85
30		23.209	17.51
40		20.042	16.41
0	14	27.032	27.11
10		25.312	21.33
20		23.842	18.04
30		23.250	17.01
40		21.923	16.30
0	28	28.419	28.00
10		26.676	25.56
20		26.299	24.15
30		24.952	20.71
40		22.923	15.91
0	90	29.028	29.11
10		27.823	26.81
20		25.655	24.74
30		24.952	21.24
40		21.750	17.63
0	120	30.210	29.92
10		29.846	28.53
20		27.215	24.89
30		22.483	20.30
40		20.479	17.33
0	180	31.302	30.15
10		30.937	29.04
20		28.208	23.78
30		22.138	21.78
40		18.934	20.67

5.0   CONCLUSIONS

From the results of the various tests performed, the following conclusions can be drawn:
1. Periwinkle shell ash had the combined acidic oxide of 50.06% and met the physical requirements for high-lime fly ash [33].
2. The slump and compacting factor decreases as the PSA content increases. This means that the concrete becomes less workable (stiff) with increase in PSA content; hence there is high demand for water to maintain the same workability level as the control.
3. The compressive strength of PSA blended cement concrete is lower than the control but there is a continuous strength development comparable with that of the control. The optimum level of PSA replacement is 10% having attained 102.22% of the design strength at 28 days.
4. The values of the Static modulus of elasticity of the control specimens (i.e. 0%PSA) is greater than those of the PSA blended cement concrete however, there was a continuous improvement with ≤20%PSA content in all the curing ages.

5. The relationship between compressive strength and static modulus of elasticity of PSA blended cement concrete up to 180 days fitted into an existing model for normal-weight concrete [29].

REFERENCES

[1] McCann, A. A. (1994), "Hydraulic concrete for pozzolanic mortars", Ueight of port city, pages: 92-97.

[2] Sahmaran, M.; Kasap, O.; Duru, K. and Yaman, I. O. (2007), "Effects of mix composition and water-cement ratio on the sulphate resistance of blended cements", Cement and Concrete Composites, pages: 159-167.

[3] Udoeyo, F. F. and Dashibil, P. U. (2002), "Sawdust Ash as Concrete Material", Journal of Materials in Civil Engineering, Vol.14 (2), pages: 173-176

[4] Zhang, H. and Malhotra, M. H. (1996), "High performance concrete incorporating rice husk ash as a supplementary cementing material", ACI Mater. Journal, Vol 93 (6), pages: 629-636.

[5] Adesanya, D. A. (2001), "The effects of thermal conductivity and chemical attack on corn cob ash blended cement", The Professional Buiders, pages: 3-10

[6] Adesanya, D. A. and Raheem, A. A. (2009), "A study of the workability and compressive strength characteristics of corn cob ash blended cement concrete", Construction and Building Materials, Vol.23 (1), pages: 311-317.

[7] Keftin, N. A. and Adole, M. A. (2006), "Properties of Concrete made with Millet Husk Ash as Partial Replacement of Cement", Journal of Environmental Sciences, Vol 10 (1), pages: 123-128.

[8] Tangchirapat, W. Jaturapitakkul, C. and Chindaprasirt, P. (2009), "Use of palm oil fuel ash as a supplementary cementitious material for producing high-strength concrete", Construction and Building Materials, Vol 23 (7), pages: 2641-2646.

[9]. Dahunsi, B. I. O. and Bamisaye, J. A. (2002), "Use of Periwinkle Shell Ash (PSA) as Partial Replacement for Cement in Concrete", Proceedings the Nigerian Materials Congress and Meeting of Nigerian Materials Research Society, Akure, Nigeria, Nov.11 – 13, pages: 184-186.

[10] Koffi, N. E. (2008), "Compressive Strength of Concrete Incorporating Periwinkle Shell Ash", Unpublished B.Sc project, University of Uyo, Nigeria.

[11] Olorunoje, G. S. and Olalusi, O. C. (2003), "Periwinkle Shell as Alternative to Coarse Aggregate in Lightweight Concrete", International Journal of Environmental Issues, Vol.1 (1), pages: 231-236.

[12] Badmus. M. A. O. Audu T. O. K. and Anyata, B. U. (2007), "Removal of Lead Ion from Industrial Wastewaters by Activated Carbon prepared from Periwinkle Shell (Typanotonus Fuscatus)", Turkish Journal of Engineering and Environmental Science, Vol 31,pages: 251-263.

[13] Mmom, P. C. and, Arokoya, S. B. (2010), "Mangrove Forest Depletion, Biodiversity Loss and Traditional Resources Management Practices in the Niger Delta, Nigeria", Research Journal of Applied Sciences, Engineering and Technology, Vol 2 (1), pages: 28-34.

[14] Powell, C. B. Hart, A. I. and Deekae, S. (1985), "Market Survey of the Periwinkle Tympanotonus fascatus in Rivers State: Sizes, Prices, Trade Routes and Exploitation levels", Proceedings of the 4th Annual Conference of the Fisheries Society of Nigeria (FISON), Fisheries Society of Nigeria, Port Harcourt, Nigeria, pages: 55-61.

[15] Jamabo, N. and Chinda, A. (2010), "Aspects of the Ecology of Tympanotonous fuscatus var fuscatus (Linnaeus,1758) in the Mangrove Swamps of the Upper Bonny River, Niger Delta, Nigeria.", Current Research Journal of Biological Sciences, Vol 2 (1), pages: 42-47.

[16] Job, O. F. Umoh, A. A. and Nsikak,S. C. (2009), "Engineering Properties of Sandcrete Blocks Containing Periwinkle Shell Ash and Ordinary Portland Cement", International Journal of Environmental Sciences, pages: 8-12.

[17] Neville, A. M.(200), "Properties of Concrete, 5th ed.", New York, Pitman.

[18] Demir, F,; Türkmen, M. and Cirak, I. (2006), "A New way for Prediction of Elastic Modulus of Normal and High Strength Concrete: Artificial Neural Networks", Proceeding of International

Symposium on Intelligent Manufacturing Systems. Dept. of Industrial engineering, Sakarya University, May, 29-31, pages: 208-215.

[19] Myers, J. J. (1999), "How to Achieve Higher Modulus of Elasticity", HPC Bridge views issue 5, Sept/Oct.

[20] Jackson, N, and Dhir, R. K. (1996), "Civil Engineering Material", London, Macmillan.

[21] Snelson, D. G.; Kinuthia, I. M.; Davies, P. A. and Chang, S. R. (2009), "Sustainable construction: Composite use of tyres and ash in concrete", Vol 29 (2), pages: 360-367.

[22] Ramezanianpour, A. A., Khani,M. M. and Ahmadibeni, G. (2009), "The Effect of Rice Husk Ash on Mechanical Properties and Durability of Sustainable Concrete", International journal of Civil Engineering, Vol 7 (2), pages: 83 – 91.

[23] Wainwright, P. J. and Tolloczko, J. J. (1986), "The Early and Later age properties of temperature OPC concrete", Second International Conference on the use of fly ash, silica fume, slag and natural Pozzolans in concrete, CANMET Vol. 2, pages: 1293 – 1321.

[24] Bhanumathidas, N., Kalidas, N. K, and Inswareb, V. (2005), "Sustainable Development through use of fly Ash, keynote paper presented at National Seminar on Building Materials and Technology for Sustainable Development", Ahmadabab, January.

[25] Nigerian Industrial Standard, NIS 444-1 (2003), "Cement- part 1: Composition, Specification and Conformity criteria for common cements", Abuja, Standard Organisation of Nigeria.

[26] British Standard Institution, BS EN 12390 part 2 (2009), "Testing hardened Concrete: Making and Curing specimens for strength tests", London, BSI.

[27] British Standard Institution, BS EN 12390 part 3 (2009), "Testing hardened Concrete: Compressive strength of test specimens", London, BSI.

[28] British Standard Institution, BS ISO 1920 part 10 (2009), "Testing of Concrete: Determination of Static Modulus of elasticity in Compression", London, BSI.

[29] Illston, J. M., (ed), (1994), "Construction materials: Their nature and behaviour 2nd ed"., London, Chapman and hall.

[30] British Standard Institution, BS 8110: Part 2 (1985), "Structural use of Concrete: Code of Practice for Special Circumstances", London, BSI.

[31] Oymael, S and Durmus, A. (2006), "Effects of Sulphates on Elastic Modulus of Concrete Samples Made from Blends of Cement with Oil Shale Ash", Oil Shale, Vol. 21 (2), pages: 125 – 134.

[32] Mehta, P. K., and Monteiro, P. J. M. (2006), "Concrete: microstructure, properties, and materials 3rd ed.", New Delhi, McGraw-Hill publishing company Ltd.

[33] American Society for Testing and Materials, ASTMC 618 (2008), "Standard Specification for Coal fly ash and raw or Calcined natural pozzolan for use in concrete", USA, West Conshohocken.

14

HYBRID PHOTOCATALYST FOR CORROSION REDUCING AND SUSTAINABLE CONCRETE CONSTRUCTION

Ranjit K. Nath[1*], M. F.M. Zain[2], Md. Rabiul Alam[2], Abdul Amir H. Kadhum[1], A.B.M.A. Kaish[2]

[1]Dept. of Chemical & Process Engineering, Universiti Kebangsaan Malaysia(UKM), Bangi, Malaysia
[2]Dept. of Civil and Structural Engineering, Universiti Kebangsaan Malaysia(UKM), Bangi, Malaysia

*Corresponding E-mail : rkn_chem@yahoo.com

ABSTRACT

Corrosion is the main problem in Reinforced concrete (RC). RC usually absorbs carbon dioxide from surrounding environment in operational period. This carbon dioxide reacts with the calcium oxide, which is already present in RC and forms a calcium carbonate an acidic compound that enhances corrosion on the surface of the reinforcement. Addition of photo catalytic materials to the RC structure during its construction phase could reduce the corrosion problem of RC materials. This material hinders calcium oxide to form acidic compound and creates a complex compound. In combination with light, this complex compound is oxidized, and residual compound will exist in or surface of RC material. In this study it is aimed to find out the proper photo catalytic material adjustable with RC material for reducing corrosion problem and enhancing oxidization process of volatile organic compounds.

Keywords: Corrosion, Carbonation, Photocatalysis, Reinforced Concrete

1.0 INTRODUCTION

Photocatalyst shows its activity in redox process. Redox reactions on the photocatalyst surface, are promoted by sunlight (or in general, weak U.V. light), and drive the oxidation of environmental pollutants [1]. The photo-induced surface hydrophilicity enhances this self-cleaning effect of the materials [2]. The addition of a photocatalyst to ordinary building materials such as concrete creates environmental friendly concrete material by which air pollution or pollution of the surface can be diminished itself [2]. For example, white concrete – whether used for precast concrete pavers or roofing – reflects much of the sun's heat and reduces the heat gain associated with dark construction materials like asphalt paving. This keeps cities cooler and reduces the need for air conditioning. It also reduces the formation of smog as chemical reactions of materials in presence of photocatalyst reduce the surface temperature of material that creates smog [1]. Photocatalyst in concrete keeps the concrete surface clean to optimize environmental benefit like degradation of organic pollutants, destruction of bacteria and viruses, decomposition of dyes or synthesis of some compounds.

When concrete or any other cement-based material in contact with the embedded reinforcing steel is carbonated, the steel surface is depassivated. Therefore, the reinforcing steel is no longer protected from corrosion. Corrosion may also commence when chloride, phosphate, moisture and oxygen gain access to the steel surface [3]-[6].One of the means of protecting steel reinforcement in concrete from chloride induced corrosion is the addition of corrosion inhibiting additives[7]-[10]. Sodium nitrite, zinc oxide, titanium dioxide monoethanol amine, diethanolamine, and triethanol amine have been used as inhibitors in concrete in different percentages for reducing chloride induced corrosion [11]. Chloride induced corrosion depends mainly on concrete pH and oxygen content. In presence of photocatalyst the corrosion problem of RC materials can be reduced [12], [13]. In this study, a hybrid photocatalyst is made from chemical mixture of ZnO and TiO$_2$ and its beneficial effects on concrete material are examined. It has also been discussed the principle and chemical combination process of hybrid photocatalyst in this study.

2.0   PRINCIPLES OF HETEROGENEOUS PHOTOCATALYSIS

Heterogeneous photocatalytic system consists of semiconductor particles (photocatalyst) which remain in close contact with a liquid or gaseous reaction medium. Exposing the catalyst to light, excited states are generated which are able to initiate subsequent processes like redox reactions and molecular transformations [14]-[16].A simplified reaction scheme of photocatalysis is shown in Fig. 1. Due to the electronic structural behavior of photocatalyst, which is characterized by a filled valence band (VB) and an empty conduction band (CB), semiconductors (metal oxides or sulfides as ZnO, CdS, TiO$_2$, Fe$_2$O$_3$, and ZnS) act as sensitizers for light-induced redox processes. The energy difference between the lowest energy level of the CB and the highest energy level of the VB is called band gap energy (Eg). It corresponds to the minimum energy of light required to make the material electrically conductive.

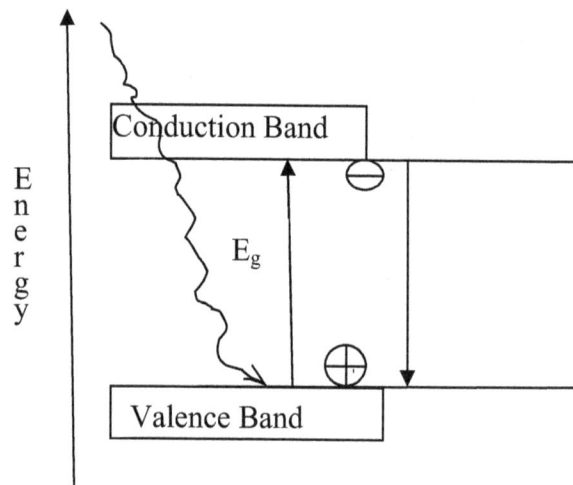

Figure 1: Operation of a photochemical excited TiO$_2$ particle

When a photon with energy of hv (h→ Plank's constant, v→ frequency of light) exceeds the energy of the band gap an electron (e-) is promoted from the valence band to the conduction band and subsequently a hole (h+) is created. In electrically conducting materials, i.e. metals, the produced charge carriers are immediately filled this hole by another electron. In semiconductors a portion of this photoexcited electron and hole together (electron-hole pairs) diffuse to the surface of the catalytic particle (electron-hole pairs are trapped at the surface) and take part in the chemical reaction with the adsorbed donor (D) or acceptor (A) molecules.

The holes can oxidize donor molecules (Equation1) whereas the conduction band electrons can reduce appropriate electron acceptor molecules (Equation2).

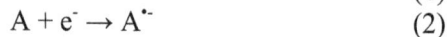

$$D + h^+ \rightarrow D^{\bullet+} \qquad (1)$$
$$A + e^- \rightarrow A^{\bullet-} \qquad (2)$$

A characteristic feature of semiconducting metal oxides is the strong oxidationpower of their holes h+. They can react in an one-electron oxidation step withwater (Equation3) to produce the highly reactive hydroxyl radical (•OH). Both the holes and the hydroxyl radicals are very powerful oxidants, which can be used to oxidize most organic contaminants.

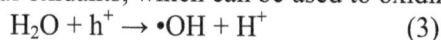

$$H_2O + h^+ \rightarrow \bullet OH + H^+ \qquad (3)$$

In general, air oxygen acts as electron acceptor (Equation4) and forms the super-oxide ion O$_2$$^{\bullet-}$ which are also highly reactive particles and are able to oxidize organic materials.

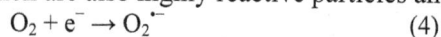

$$O_2 + e^- \rightarrow O_2^{\bullet-} \qquad (4)$$

3.0   COMBINATION OF ZnO AND TiO₂ AS HYBRID PHOTOCATALYST

The ZnO is found in the form of hexagonal crystal or as a white powder. It also occurs in nature as mineral zincates. The Wurtzite structure of ZnO is a wide band gap semiconductor with band gap energy of near about 3.2ev, which corresponds to electromagnetic radiation in the UV (ultraviolate) region. The top of valence band is composed of 2p orbital from the O^{2-} atoms while the bottom of the conduction band is composed of 4s orbital from the Zn^{2+} atom.

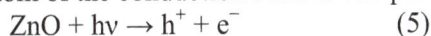

$$ZnO + hv \rightarrow h^+ + e^- \qquad (5)$$

Therefore, the creation of electron hole pairs is through a ligand to metal charge transfer of an electron from the UV on oxygen to the conduction band on zinc.

Figure 2: Zinc oxide powder   **Figure 3:** unit cell of ZnO

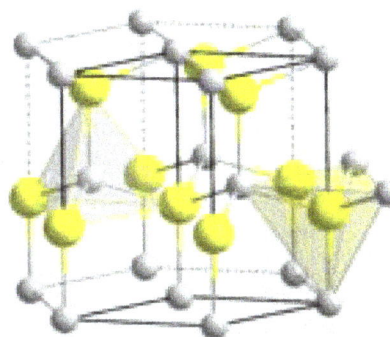

The reaction rate of the photocatalyst is surface area dependent, as its electron- hole transfer takes place on the surface. Therefore, the photocatalyst morphology is an important property in determining the effectiveness of the photocatalyst [17]. When the surface-to-volume ratio increases, the rate of electron-hole pairs transfer from the semiconductor to the absorbed molecule also increases due to the ability of the ZnO to absorb more molecules on the surface.

TiO₂ exists in three crystalline forms: rutile, anatase, and brookite (Fujishima et al.)[18].Titanium dioxide, particularly in the anatase form, is a photocatalyst under ultraviolet (UV) light. Recently, it has been found that titanium dioxide, when spiked with nitrogen ions or doped with metal oxide like tungsten trioxide, works as a photocatalyst under either visible or UV light[19]. The strong oxidative potential of the positive holes oxidizes water and creates hydroxyl radicals. It can also oxidize oxygen or organic materials directly. Titanium dioxide is thus added to paints, cements, windows, tiles, or other products for its sterilizing, deodorizing and anti-fouling properties and is used as a hydrolysis catalyst.

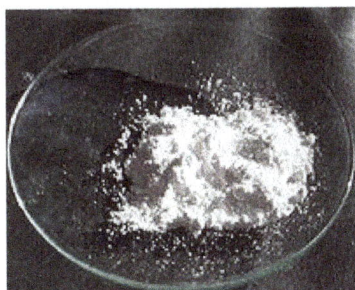

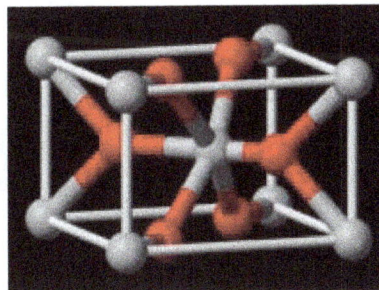

Figure 4: Crystals of TiO₂   **Figure 5:** Titanium dioxide powder   **Figure 6:** unit cell of TiO₂

Generally, titanium dioxide is a semiconducting material which can be chemically activated by light. Under the influence of light the material tends to decompose organic materials in presence of photocatalyst. This effect leads to the well-known phenomenon of"paint chalking", where the organic components of the paint are decomposed asresult of photocatalytic processes [20]. Fujishima and Honda discovered the photocatalytic splitting of water onTiO₂ electrodes

[21]. This event marked the beginning of a new era in heterogeneous photocatalysis. Although TiO_2 absorbs only approx. 5 % of the solar light reaching the surface of the earth, it is the best investigated semiconductor in the field of chemical conversion and storage of solar energy [21].

The conduction band of TiO_2 is composed of 3d, 4s and 4p orbitals from Ti^{4+} atoms, with the 3d orbitals being the majority of the lower portion of the conduction band. Band gap energy of this material is Eg = 3.2eV. If this material is irradiated with photons of the energy > 3.2eV (wavelength λ < 388 nm), the band gap increases and an electron is promoted from the valence to the conduction band. Consequently, charge-carrier is formed in its primary process (Equation 6).

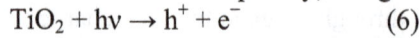

$$TiO_2 + h\nu \rightarrow h^+ + e^- \qquad (6)$$

The ability of a semiconductor to undergo photo induced electron transfer to adsorbed particles is governed by the band energy positions of the semiconductor and the redox potentials of the adsorbents [22], [23]. The relevant potential level of the acceptor species is thermodynamically required to be below the conduction bandof the semiconductor. Otherwise, the potential level of the donor is required to be above the valence band position of the semiconductor in order to donate an electron to the empty hole.

From the theoretical description of the localized atomic-like Ti 3d state is induced by Zn^{2+} ions. It is seen that hybrid photocatalyst provides satisfactory results in increasing surface area, thermal stability and surface acidity.

4.0   EXPERIMENTAL   DETERMINATION   OF p^H OF   HYBRID PHOTOCATALYST AND CEMENTATIONS MATERIALS

4.1   DETERMINATION OF APPROPRIATE PROPORTION OF ZnO AND TiO_2 AS HYBRID PHOTOCATALYST

Highly pure TiO_2 and ZnO were used for measuring the appropriate proportion of the mixture for hybrid photocatalyst. At first 16mg cement was mixed with 100ml distilled water for maintaining 1N catalyst to cement ratio 1:100. Then 1N solution of TiO_2 and1N solution of ZnO were prepared separately. Thereafter, pH of individual solution was measured using BUTECH-810 pH meter. Solution of both photocatalystes was mixed at different ratios(As indicated in graph) and pH of the mixture was measured using the same pH meter. pH value for different mixing proportion of TiO_2 and ZnO are shown in Fig.-7.

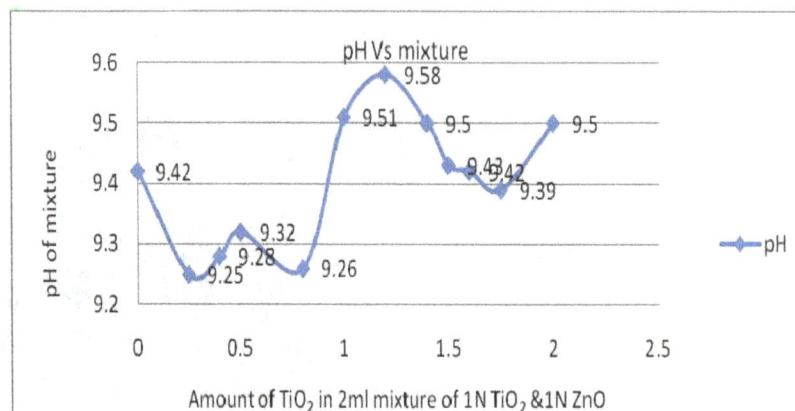

Figure 7: p^H for different mixing proportion of 1N TiO_2 and1N ZnO in mixture (2ml)

From this figure, it is seen that pH value is maximum at a mix proportion of TiO_2 and ZnO of 1.2: 0.8. This pH value is also found to be greater than the pH value of individual photocatalyst.

4.2   DETERMINATION OF p^H OF CEMENTATIONS MATERIALS IN PRESENCE OF HYBRID PHOTOCATALYST

In order to determine the pH of cementations materials in presence of hybrid photocatalyst cement solution was made first by mixing 16mg cement into 100ml distilled water for maintaining 1N catalyst to cement ratio 1:100. Thereafter, 1.2 ml 1N TiO_2 and 0.8ml 1N ZnO were added to the 4ml cement solution (taken from the instant mixture of cement solution) for obtaining solution of cementations material in presence of photocatalyst. pH of this solution was measured instantly using the same pH meter used to measure pH of photocatalyst pH value of cementations material in presence of individual photocatalyst was also measured using the same pH meter in this study. The pH values obtained from this study are given in Table 1.

Table 1: p^H of photocatalyst in presence of cement

Component	p^H
Cement+ Hybrid photocatalyst	10.44
Cement+ ZnO	10.39
Cement+ TiO$_2$	10.26

From Table 1 it is seen that the mixture of 1N TiO_2 and 1N ZnO at a ratio of 1.2:0.8 with cement solution gives the higher pH (10.44)value than that of the individual mixture of 1N TiO_2 (pH 10.39)and 1N ZnO (pH 10.26) with cement solution. Higher pH value of cementations materials in presence of hybrid photocatalyst reduces the corrosion problem of RC materials.

5.0   ADVANTAGES OF HYBRID PHOTOCATALYST

The heterogeneous photocatalytic oxidation with ZnO and TiO_2 could make it competitive with respect to other processes oxidizing contaminants. It has the following advantages:
  ➢ Low-cost material is used as photocatalyst.
  ➢ The reaction is quite fast at mild operating conditions (room temperature and atmospheric pressure).
  ➢ No chemical reactants are used and no side reactions are produced..

6.0   APPLICATION OF ZnO AND TiO$_2$ AS HYBRID PHOTOCATALYST

Photocatalyst is needed for a cleaner environment and a better quality of life that leads to thoughts of a more eco-compatable use of light. In this context photochemistry applied to construction materials that could provide a very interesting solution [24]. It could become an integral part of the strategies adopted for reducing environmental pollution through the use of construction materials containing photocatalysts [25].

Hybrid photocatalyst of ZnO and TiO_2, which creates local energy levels within the band gap of the photocatalyst, with corresponding absorption bands lying in the visible spectral range [26]. It is assumed that the photoexcitation of such hybrid photocatalyst to the generation of free charge carriers to initiate surface chemical processes. However, the efficiency of such systems under visible light strongly depended on the preparation method used and ratio of photocatalysts.

In figure 8, the main areas of activity in hybrid photocatalysis are shown. As already mentioned, in the last 10 years photocatalysis has become more and more attractive for the industry regarding the development of technologies for purification of water and air

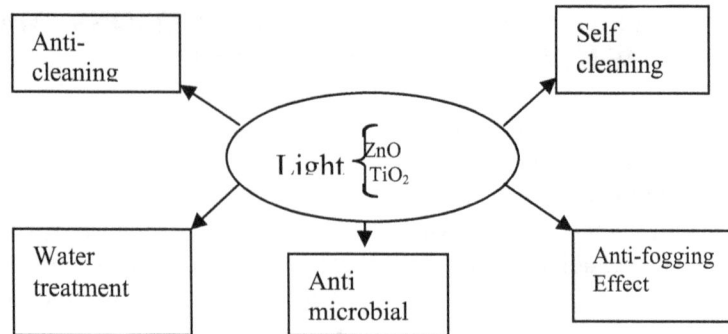

Figure 8: Major areas of activity in mixed photocatalysis

7.0   CONCLUSIONS

From the experimental study of p^H measurement of cementitious material in the presence of hybrid photocatalyst (made by ZnO and TiO_2) it could be concluded that hybrid photocatalyst could help to reduce corrosion effect of R.C. concrete as its pH value are found to be higher than the pH value of cementation material with individual photocatalyst. Since hybrid photocatalyst (ZnO and TiO_2) has large surface area, therefore, it exhibits a good photo degradation property. Hybrid photocatalyst could also help to absorb more carbon dioxide into the concrete from atmosphere due to creation of lower activation energy in their process. This phenomenon produces sustainable environment for human being. Though this study is in developing stage it needs to consider employing many of the options that might improve the sustainability of this material, and continue to gain knowledge and expertise in this area.

ACKNOWLEDGEMENT

At first, the authors wish to express gratitude to Almighty God for beginning this research. The authors would like to express special thanks to Universiti Kebangsaan Malaysia; and Faculty of Engineering and Built Environment for allocating funds for the research. The first author expresses honest appreciation to Chittagong University of Engineering and Technology (CUET), Chittagong, Bangladesh for providing him leave for the research.

REFERENCES

[1] Fujishima. A, Hashimotoand. K and Watanabe. T (1999), "TiO2 Photocatalysis: Fundamentals and Application", Tokyo, BKC, pages: 123-129.
[2] Wang. R, Hashimoto. R, Fujishima. A, Chikuni. M, Kojima. E, Kitamur. K, Shimohigoshi. M and Watanabe. T (1997), "Light-induced amphiphilic surfaces", Nature, vol. 338, pages: 431-432.
[3] Cigna. R, Familiari. G, Gianetti. F, and. Proverbio. E (1994), "Corrosion and Corrosion Protection of Steel in Concrete (ed, R. N. Swamy)", Sheffield, UK, page: 848.
[4] Chin. D (1987), "Concrete and Chloride", ACI SP 102, page: 49.
[5] El-Jazairi. B and Berke. N. S (1990), "Corrosion of Reinforcement in Concrete (eds. C. L. Page, K. W. J. Treadaway, and P. B. Bamforth)", Elsevier Applied Science, page: 571.
[6] Raharinaivo. A, Bouzanne. M and Malric. B (1997), "Influence of Concrete Aging on the Effectiveness of Monofluoro phosphate for Mitigating the Corrosion of Embedded Steel", page: 585.
[7] Chin. D (1989), "Concrete and Corrosion", ACI SP 108, page: 38.
[8] Miksic. B. A, Chandler. C, Kharshan. M, Furman. A, Rudman. B and Gelner. L (1997), "Corrosion inhibitor for use in reinforced concrete structures", U.S, US 5,597, pages: 514- 28.

[9] Elsener. B, Büchler. M and Böhni. H (1997), "Corrosion Inhibitors for Steel in Concrete", page: 469.

[10] Elsener. B and Böhni. H (1998), "Corrosion Inhibitors for Reinforcement in Concrete", page: 412.

[11] Ha-Won Song and Velu Saraswathy (2006), "Analysis of Corrosion Resistance Behavior of Inhibitors in Concrete using Electrochemical Techniques", METALS AND MATERIALS International, Vol 12(4), pages: 323~329

[12] Cigna. R, Familiari. G, Gianetti. F, Proverbio. E (1994), "Corrosion and Corrosion Protection of Steel in Concrete (ed, R. N. Swamy)", Sheffield, UK, page: 848.

[13] El-Jazairi. B and Berke. N. S (1990), "Corrosion of Reinforcement in Concrete (eds. C. L. Page, K. W. J. Treadaway, and P. B. Bamforth)", Elsevier Applied Science, page: 571.

[14] Gopidas. K. R, Kamat. P. V, (1993), "Photoinduced charge transfer processes in ultrasmall semiconductor clusters, Photophysical properties of CdS clusters in fion membrane", Proc. Ind. Acad. Sci (Chem. Sci); 105, pages: 505-512.

[15] Sant. P. A, Kamat. P. V (2002), "Inter-particle electron transfer between size quantized CdS and TiO2 semiconductor nano clusters", Phys. Chem. Chem. Phys; 4, pages: 198-203.

[16] Swamy. R. N "The alkali- silica reaction in concrete", ISBN 0-216-92691-2.

[17] Lizama. C, Freer. J, Baeza. J, and Mansilla. H.D (2002), "Optimized photodegradation of reactive Blue 19 on TiO2 and ZnO suspensions", Catal. Today, Vol. 76, pages: 235-246.

[18] Fujishima. A., Hashimoto. K., Watanabe. T. (1999), "TiO2, photocatalysis, fundamentals and applications", BKC, Inc., page: 176.

[19] Kurtoglu. M. E, Longenbach. T, Gogotsi. Y (2011), "Preventing Sodium Poisoning of Photocatalytic TiO2 Films on Glass by Metal Doping", International Journal of Applied Glass Science2 (2): 108–116. doi:10.1111/j.2041-1294.2011.00040.x.

[20] Vinodgopal. K, Bedja. I, Kamat. P.V(1996), "Nanostructured semiconductor films for photocatalysis. photoelectrochemical behavior of SnO2/TiO2 coupled systems and its role in photocatalytic degradation of a textile azo dye", Chem. Mater; 8(8): 2180.

[21] Fujishima. A, Honda. K (1982), "Electrochemical Photolysis of Water at a Semiconductor Electrode", Nature, 238, 37.

[22] Linsebigler. A L, Lu. G and Yates. J. T (1995), "Photocatalysis on TiO2 surfaces: principles, mechanisms and selected results", Chem. Rev., 95, pages: 735-758.

[23] Hashimoto. K, Irie. H and Fujishima. A (2005), "TiO2 Photocatalysis: A Historical Overview and Future Prospects", Jap. J. Appl. Phys. Part 2: Lett., 44 , pages: 8269-8285.

[24] Ranjit K. Nath, M. F.M .Zain, Rabiul Alam, Abdul Amir H. Kadhum (2012), " A review on the effect of carbonation and accelerated carbonation of cementitious materials and its consequence in waste treatment", Journal of Applied Sciences Research, 8(5): 2473-2483.

[25] Ranjit K. Nath, M. F.M .Zain, Abdul Amir H. Kadhum (2012), "New Material LiNbO3 for photocatalytically improvement of indoor air - An overview". Advances in Natural and Applied Sciences, 6(7): 1030-1035, ISSN 1995-0772

[26] Ballari, M.M., Yu, Q.L., Brouwers, H.J.H., (2010), "Experimental study of the NO and NO2 degradation by photocatalytically active concrete", Catalysis Today, 151, 231-239.

15

EVALUATION OF THE CONTRIBUTION OF CONSTRUCTION PROFESSIONALS IN BUDGETING FOR INFRASTRUCTURE DEVELOPMENT IN NIGERIA

Opawole, A., Jagboro, G.O. Babalola O. and *Babatunde S.O

Department of Quantity Surveying, Obafemi Awolowo University, Ile-Ife, Nigeria

*Corresponding E-mail : Sholly_intl@yahoo.com

ABSTRACT

Researchers are of the opinion that the low implementation of public financed infrastructure projects in Nigeria could be correlated to the level of involvement of construction professionals in the budgeting process at macro-level. Though this assertion presently lacks empirical justification, the objective of this study seeks to quantitatively establish this linkage. In order to achieve this, fourteen (14) core budgeting and procurement processes were identified in literature. Respondents involved in the study were architects, quantity surveyors, builders, town planners, estate surveyors, engineers (civil, mechanical and electrical), accountants and economists in the public service of Osun state. The fact that infrastructure financing depends majorly on budgetary financing in Osun state provided the justification for choice of the State for the study. Descriptive and inferential statistics were adopted to analyse the data collected. The study indicates inadequate contribution of construction professionals in activities involving post-budgetary activities and only progressive trend in pre-budgetary process especially technical and cost evaluation of infrastructure projects and review and approval of budgets for infrastructure projects. Moreover, budgeting process for infrastructure development in Nigeria indicated that majority of projects budgeted for execution lack adequate technical evaluation and cost assessment as a result of inadequate professional involvement. This could be adduced to be a significant problem of implementation of public financed infrastructure projects in Nigeria. The study provides information on key areas where public policy makers can appropriate construction professionals' inputs to prepare realistic budget for infrastructure development in a developing economy.

Keywords: *Infrastructure, budgeting, construction professionals*

1.0   INTRODUCTION

Infrastructure has come increasingly to be recognized as a very strong parameter and index for measuring a nation's global competitiveness. Infrastructure such as road, electricity, water supply, hospitals, telecommunication and security system among others facilitate agricultural, industrial and commercial production; render social services; and maintain the security of a community [1]. Infrastructure procurement is basically through public financing in Nigeria [2]. This is due to the low level participation of the private sector in infrastructure development and the present embryonic state of Private-Public Partnership (PPP) financing initiative. However, growing private participations have been recorded in the area of transportation, waste management and commerce [3, 4], which were executed in form of joint venture [5] and concession variants arrangements [6]. This situation makes the public policy makers and the construction industry the major actors in infrastructure development in Nigeria. Public financing essentially relates to infrastructure procurement through budgetary allocation [7]. According to [8], capital expenditure staggers between 65–70% in the annual budgets of the three tiers of government in Nigeria, and infrastructure is often responsible for about 50% of the capital expenditure. Gray and Larson [9] identified infrastructure procurement as subject to sequential stages of identification, definition, planning, execution, and delivery, and these stages as crucial to success of infrastructure construction. Notwithstanding, sensitive stages, especially, identification, definition, planning, and budgeting, for infrastructure sector at macro level have been criticized to be dominated by the executive arm of the government with minimum input of the construction professionals [10, 11]. Oforeh [10] asserted that the policy makers who plan for infrastructure development in both the national and state budgets lack adequate knowledge of the complex technological processes of construction and the cost characteristics of infrastructure constructions. The study identified shortfall in budgetary allocation for infrastructure as the inability of the policy makers to plan adequately for the sector. This is evident from the present level of stock of infrastructure in Nigeria.

Aside the criticism that annual budgets are poorly planned in Nigeria, implementation of budget in Nigeria is identified to be characterized by fiscal indiscipline and bureaucracy, resulting most often in abandonment of projects [12, 10]. More significant problems of infrastructure development in Nigeria include insensitive and disjointed government policies; wrong application of procurement methods; deficient procurement procedures; and dominance of foreign technical manpower to the detriment of indigenous manpower [13]. As a result of these scenarios, infrastructure development in Nigeria do not record success as anticipated. In the light of the prominent role of infrastructure in improving the standard of living and economic growth, and the present scenarios in infrastructure development in Nigeria, this study focuses on appraising infrastructure development in Nigeria with specific emphasis on the contribution of construction professionals to the budgeting process for the infrastructure sector.

2.0 LITERATURE REVIEW

Majority of studies on infrastructure development in Nigeria were limited to the socio-economic aspect. Notable among these is [14] who studied investment in telecommunication infrastructure. The study showed that a US$1 invested in telecommunication infrastructure generates an economic return of US$6 by way of its impact on local employment and general economic growth. Studies by [15] and [16] related primarily to social and economic impact of infrastructure on national and regional economic growth and development. Findings from these studies similarly established positive correlation between investment in infrastructure and more rapid economic growth and decline in poverty.

Study by [17] relating to infrastructure development in the educational sector identified the financing of educational infrastructural projects as substantially through public budgetary allocation and concluded that poor funding is a major challenge in the development of educational sector in Nigeria. The issue of procurement and financing attracted the attention of [18] which studied on financing strategies for infrastructure development in Nigeria. The study which is an extension and extrapolation of [17] limited work on the educational sector, identified infrastructure financing in Nigeria as substantially through public budgetary allocation. It concluded that financing is one of the most fundamental issues that is germane to success of infrastructure development.

Olayiwola and Adeleye [19] study on the challenges and problems of rural infrastructural development in Nigeria highlighted the concept of rural infrastructural planning and examined the Nigerian rural infrastructural policies over the years 1960-1990. The major problems and challenges posed by the various rural infrastructural development identified include the lack of spatial focus in rural development planning; lack of perceptual focus in the development plans; restriction of means of rural infrastructural provision to public funding; and lack of action and appropriate institutional arrangements for the execution of rural infrastructural programmes.

Despite the preponderance of research effort, it appears no attempt has been made by researchers to investigate the fundamental reasons why infrastructure projects undertaken through public budgetary allocations are in most cases not fully and effectively implemented in Nigeria. This according to [11] and [10] was asserted to the low level of involvement of construction professionals in the budgeting process for the infrastructure sector. While this assertion presently lacks empirical justification, the objective of this study was to investigate a quantitative linkage between construction professionals and the budgeting process in the execution of infrastructure projects in Osun state, South western Nigeria.

Robert and Lynch [20] defined budget as an estimate of the government income and expenditure which occurs in four phases of process, that is, policy planning and resource analysis, policy formulation, policy execution and evaluating the entire process and system. According to Olufidipe [12], budgeting process include provision of the plan of action for implementing government programmes; preparation of the strategies for implementing the plan; issuance of call circulars to executive; preparation of budgets estimate; review and adjustment of the budget/estimates; preparation of consolidated estimate of revenue and expenditure (CERE) and its presentation to the to the legislators in form of appropriation bill and to the executive.

According to Robert and Lynch [20], the political executives see the budgeting process as a political event conducted in the political arena for political advantage while economic analysts view budgeting as a matter of allocating resources in terms of opportunity cost. The United State Agency for International Development [21] report identified the process of budgeting as a significant factor influencing infrastructure projects implementation in Nigeria. The study reported the major problems of the budgeting process as lack of political will and commitment to abide by stipulated rules and budget guidelines; inability to develop a macro-economic framework for budget formulation; ambiguities in the roles of various agencies involved in the formulation and monitoring of the budget; periodic changing of budget line items classifications; lack of coordination in the disbursement of funds after budget approval and slow budget process fraught with errors.

Oforeh [10] and Mogbo [11] assessed the stages involved in infrastructure development, especially, identification, definition, planning, and budgeting, at macro-level as being dominated by the executive arm of the government with minimum input of the construction professionals in Nigeria. Oforeh [10] further asserted that the policy makers who plan for infrastructure development in both the national and state budgets lack adequate knowledge of the complex technological processes of construction and the cost characteristics of infrastructure constructions. The study further identified shortfall in budgetary allocation to infrastructure as the inability of the policy makers to plan adequately for the sector and thus consequently have impacted negatively on implementation of infrastructural projects in Nigeria. Upon this theoretical background, this study assessed the quantitative linkage between construction professionals and the budgeting process in the execution of infrastructure projects in Osun state, South western Nigeria.

3.0   METHODOLOGY

The data for this study were collected through structured questionnaire administered on building industry and other allied financial experts that are involved in the execution of infrastructural projects. These respondents include architects, quantity surveyors, builders, town planners, estate surveyors, engineers (civil, mechanical and electrical), accountants and economists in the public service of Osun state. The issues included in the study are related to the assertion raised by [11] and [10] on the significance of contribution of construction professionals to budgeting for infrastructure development at macro-level in Nigeria. The choice of Osun state, in the South-western region of Nigeria, as the study area was justified by fact that infrastructure development in the state depends substantially on budgetary financing [22]. A total of seventy-two (72) questionnaire were completed by 6 architects, 4 quantity surveyors, 6 town planners, 5 estate surveyors, 4 builders, 21 engineers (mechanical, civil, and electrical) and 26 economists/accountants which represented a response rate of 70% of total 106 questionnaires administered. The distribution of the respondents is shown in figure 1. The questionnaire was of two parts. The first part identified the demographic features of the respondents and the second part relates to involvement of the professionals in infrastructure budgeting process. The respondents were asked to score the extent of their involvement in the budgeting for infrastructure sector in the State on the scale of 0-100% where 0 represents lowest ranking and 100 representing highest ranking. The results were presented in tables 1 and 2. The data obtained was analyzed by descriptive and inferential statistics.

4.0   RESULTS AND DISCUSSIONS

Figure 1 shows the percentage representation of the respondents as 8.3% for architects, 8.3% for town planners, 5.6% for builders, 5.6% for quantity surveyors, 29.2% for engineers, 6.9% for estate surveyors and 36.1% for economists/accountants. Moreover, the respondents were restricted to professionals with official cadre ranging between principal to directors. These are the officials that were purported to have been involved in government decision making process including budgeting and stand the position to supply reliable data for the study.

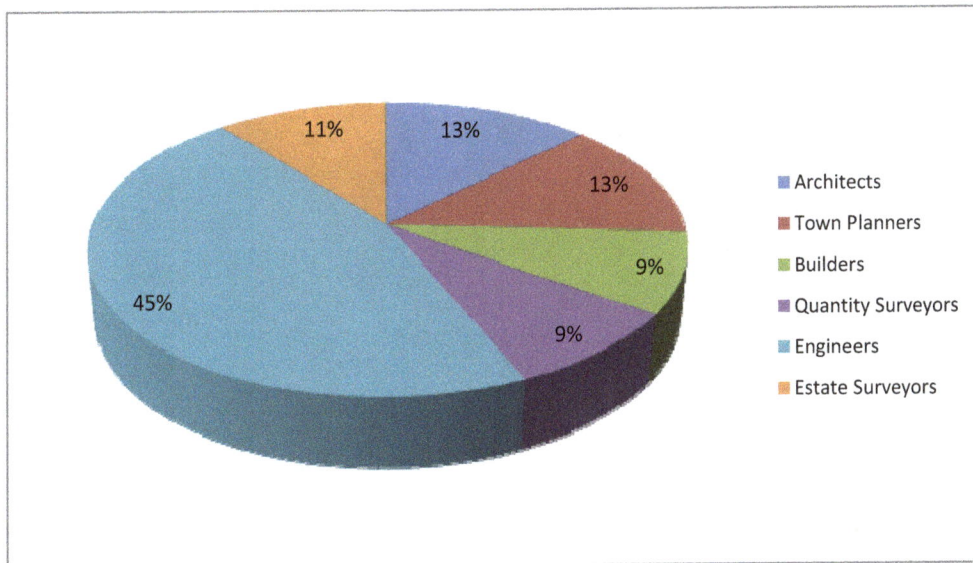

Figure 1: Respondents Classification by Profession

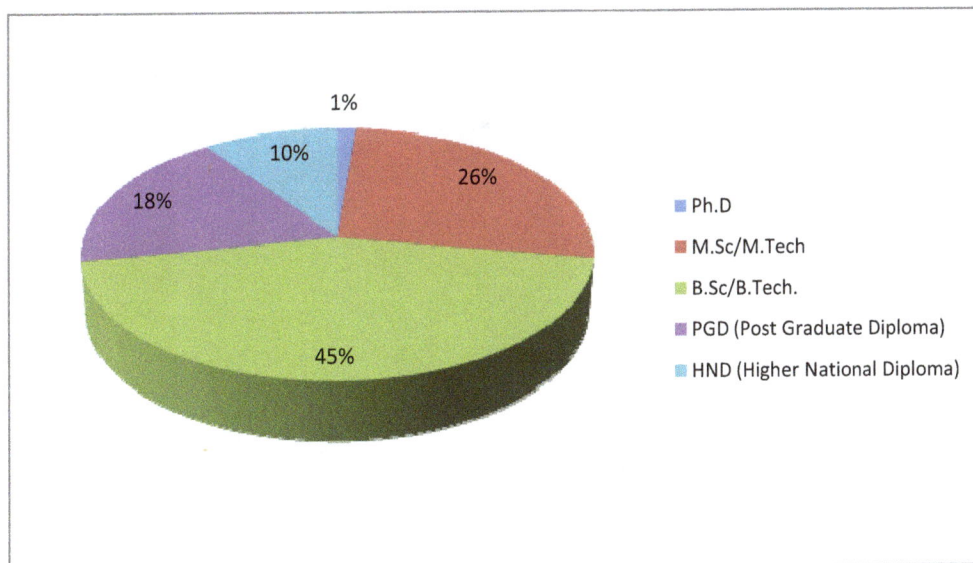

Figure 2: Academic Qualification of the Respondents

Figure 2 shows that 26.4% of the respondents are holders of Master of Science or Masters of Technology; 44.5% are holders of Bachelor of Science or Bachelor of Technology; 18.1% obtained Post Graduate Diploma (PGD); 9.7% holds Higher National Diploma (HND); and 1.4% holds Doctor of Philosophy. The results show that all the respondents possess the minimum registration qualification of their various professional bodies in Nigeria and are of adequate academic training to supply reliable data for this study.

In Figure 3 the mean industry work experience is estimated as 14 years, which represents the working experience of about 52% of the respondents. With this average working experience of fourteen years, respondents are deemed experience enough to supply reliable date for this study.

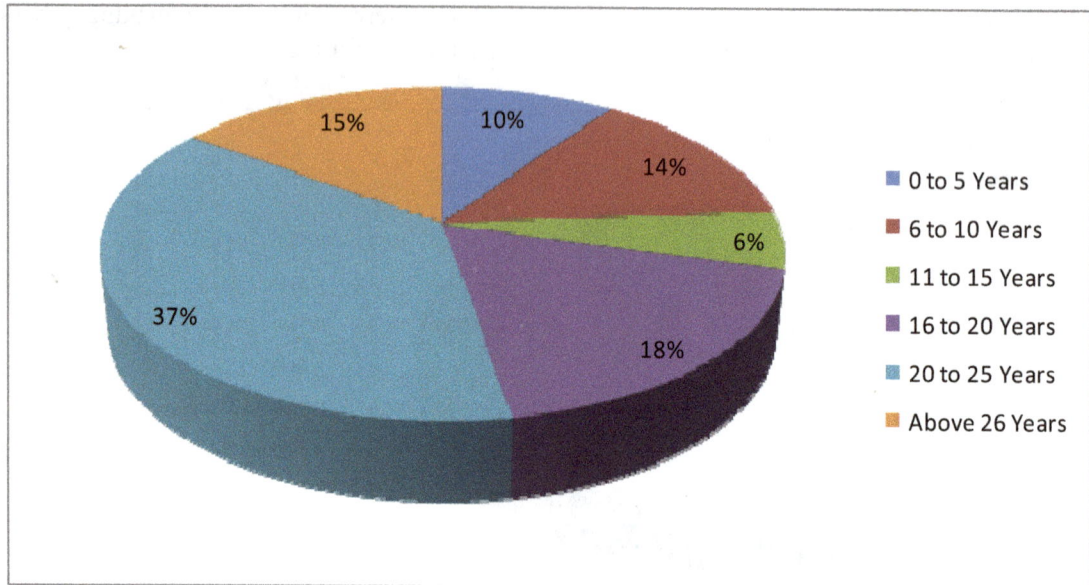

Figure 3: Working Experience of Respondents

Figure 4 shows the professional qualification of the respondents. The result shows that the respondents are either associate or corporate members of the various professional bodies or posses some other professional qualification. This shows that the respondents are in the position to supply reliable data for the research.

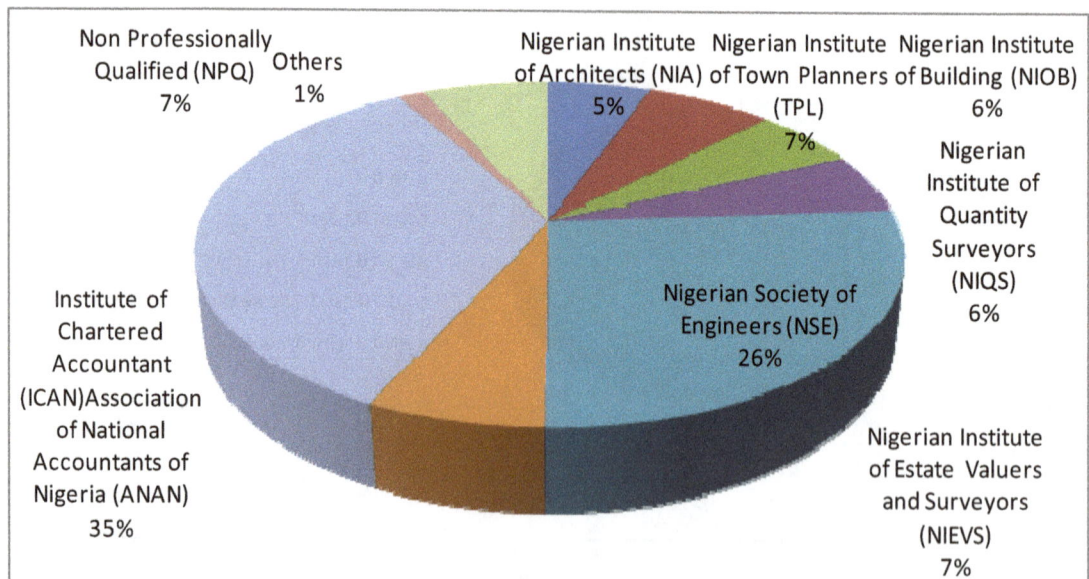

Figure 4: Professional Qualification of the Respondents

Tables 1 and 2 show the quantitative contribution of construction professionals in budgeting process for infrastructure development. For the Architects, the tables reveal the highest contribution of 75.50% which are equal for participation in projects execution, monitoring and evaluation process; and projects cost monitoring and control activities. Next to these are participation in projects tender and selection process, drafting of projects contractual agreement/documentation; preparation of budget estimate for infrastructure projects; and finalization of the draft budget for overall infrastructure sector with percentage contribution of 72.17%, 63.75, and 55.42% and 52.08% respectively. The least on the ranking are part of budget committee to identify the state's infrastructure needs; participation in developing of budget strategy to accommodate the infrastructure needs; preparation of budget circular which gives guidelines for the preparation of infrastructure sector budgets; with respective percentage contribution of 38.67%, 38.75%, and 40.33%.

In the case of Town Planners, the percentage contribution ranges between 20.25% - 43.83%. Preparation of budget estimate for infrastructure projects; part of budget committee to identify the state's infrastructure needs; and participation in review and approval of budget estimate for infrastructure projects which received the highest ranking of 43.83%, 35.42%, and 33.75% respectively still fall below average. The tables reveal the least contribution in project budget auditing activity (20.25%); preparation of the macro-economic framework (22.00%) and drafting of projects contractual agreement/documentation (22.00%).

The tables reveal a percentage contribution of 78.00% for participation in projects execution, monitoring and evaluation process; equal percentage contribution of 65.50% for drafting of projects contractual agreement/documentation; participation in projects tender and selection process; and projects cost monitoring and control activities; 58.00% for preparation of budget estimate for infrastructure projects; and 55.50% for project budget auditing activity for quantity surveyors.

Table 1: Contribution of Construction Professionals in Budgeting for Infrastructure Development in Osun State

Process Involved	Construction Professionals (%)							Financial Administrators (%)
	Arch.	TPL	Bldr.	Quantity Surveyor	Engr.	Estate Surveyor	Aggreg.	Accountant/ Economist
Preparation of the macroeconomic framework for development of infrastructure sector	50.42	22.00	58.00	30.38	42.57	45.50	41.51	43.52
Preparation of a budget circular which gives guidelines for the preparation of infrastructure sector budgets	40.33	32.00	55.50	15.38	41.10	53.50	40.61	50.08
Part of budget committee to identify the state's infrastructure needs	38.67	35.42	58.00	27.88	48.29	43.50	42.37	50.06
Participation in pre-budgetary technical and cost evaluation of infrastructure projects	48.75	32.08	60.50	48.00	51.19	59.50	50.68	53.17

Process	Arc	TPL	Bldr	Engr	ES	Agg		Aggreg
Participation in developing of budget strategy to accommodate the infrastructure needs	38.75	30.42	68.00	43.00	38.79	55.50	43.07	45.10
Preparation of budget estimate for infrastructure projects	55.42	43.83	65.50	58.00	56.90	53.50	54.83	55.48
Participation in review and approval of budget estimate for infrastructure projects	45.33	33.75	63.00	48.00	53.07	53.50	50.01	50.85
Finalization of the draft budget for overall infrastructure sector	52.08	25.42	60.50	30.38	42.57	51.50	43.04	44.25
Participation in preparation of implementation plan of budget for infrastructure sector	45.42	23.75	60.50	50.38	42.10	53.50	42.84	47.75
Drafting of projects contractual agreement/documentation	63.75	22.00	78.00	65.50	50.17	31.30	48.00	52.35
Participation in projects tender and selection process	72.17	30.33	78.00	65.50	53.07	43.50	54.80	51.19
Participation in projects execution, monitoring and evaluation process	75.50	23.67	75.50	78.00	67.86	47.40	64.37	49.65
Projects cost monitoring and control activities	75.50	25.25	75.50	65.50	61.67	45.50	58.50	50.40
Project budget auditing activity	47.08	20.25	75.50	55.50	43.02	47.50	44.53	47.35

Source: Author's Field Survey (2010)

Legend: Arc =Architect; TPL= Town Planner; Bldr = Builder; Engr = Engineer; Aggreg =Aggregate

Table 2: Ranking of Contribution of Construction Professionals in Budgeting for Infrastructure Development

Process Involved	Construction Professionals (%)							Financial Administrators (%)
	Arch	TPL	Bldr	QS	Engr	E S	Aggregate	Accountant/ Economist
Preparation of the macroeconomic framework for development of infrastructure sector	7	12	12	11	10	10	14	14
Preparation of a budget circular which gives guidelines for the preparation of infrastructure sector budgets	12	7	14	14	12	3	11	7

Part of budget committee to identify the state's infrastructure needs	14	2	12	13	8	12	9	8
Participation in pre-budgetary technical and cost evaluation of infrastructure projects	8	6	9	8	6	1	5	2
Participation in developing of budget strategy to accommodate the infrastructure needs	13	5	6	10	14	2	13	12
Preparation of budget estimate for infrastructure projects	5	1	7	5	3	3	3	1
Participation in review and approval of budget estimate for infrastructure projects	11	3	8	8	4	3	6	5
Finalization of the draft budget for overall infrastructure sector	6	8	9	11	10	7	12	13
Participation in preparation of implementation plan of budget for infrastructure sector	10	10	9	7	12	3	10	10
Drafting of projects contractual agreement/documentation	4	12	1	2	7	14	7	3
Participation in projects tender and selection process	3	4	1	2	4	12	4	4
Participation in projects execution, monitoring and evaluation process	1	11	2	1	1	9	1	9
Projects cost monitoring and control activities	1	9	2	2	2	10	2	6
Project budget auditing activity	9	14	2	6	9	8	8	11

Source: Author's Field Survey (2010)

Legend: Arc =Architect; TPL= Town Planner; Bldr = Builder; Engr = Engineer; Aggreg =Aggregate

On the lower scale of the ranking of quantity surveyors' contribution, the tables reveal 15.38% for preparation of a budget circular which gives guidelines for the preparation of infrastructure sector budgets; 27.88% for participation in identifying the state's infrastructure needs; equal percentage contribution of 30.38% for both preparation of the macro-economic framework for development of infrastructure sector; and finalization of the draft budget for overall infrastructure sector.

The contribution of engineers in projects execution, monitoring and evaluation process ranks first with the percentage contribution of 67.86%. Following this closely are projects cost monitoring and control activities with percentage contribution of 61.67%; preparation of budget estimate for infrastructure projects with percentage contribution of 56.90%; and participation in review and approval of budget estimate for infrastructure projects; and participation in projects tender and selection process with equal percentage contribution of 53.07%. The contribution was least in participation in developing of budget strategy to accommodate the infrastructure needs with mean percentage contribution of 38.79%; participation in preparation of implementation plan of budget for infrastructure sector; and preparation of a budget circular which gives guidelines for the preparation of infrastructure sector budgets both with percentage contribution of 42.10%; and finalization of the draft budget for overall infrastructure sector and preparation of the macroeconomic framework for development of infrastructure sector both also with percentage contribution of 42.57%.

Builders' contribution in drafting of projects contractual agreement/documentation; and projects tender and selection process ranked first with the mean percentage of 78.00% for both, followed by projects' cost monitoring and control activities; projects budget auditing activity; and participation in projects execution, monitoring and evaluation process each with percentage contribution of 75.50%. Participation in developing of budget strategy to accommodate the infrastructure needs was ranked sixth with percentage contribution of 68.00%. Preparation of a budget circular which gives guidelines for the preparation of infrastructure sector budgets ranked lowest with 55.50% followed by preparation of the macro-economic framework for development of infrastructure sector and part of budget committee to identify the state's infrastructure needs both with equal ranking of 58.00%. On overall, the rankings reveal a range of 55.50% - 78.00% for builders.

The contribution of estate surveyors shows participation in technical and cost evaluation of infrastructure projects which is ranked first with the percentage contribution of 59.50%. Following this closely are developing of budget strategy to accommodate the infrastructure needs with percentage contribution of 55.50%; participation in review and approval of budget estimate for infrastructure projects and preparation of implementation plan of budget; preparation of a budget circular which gives guidelines for the preparation of infrastructure sector budgets; and preparation of budget estimate for infrastructure projects with equal percentage contribution of 53.50%. Low contributions were obtained in drafting of projects contractual agreement/documentation (43.50%); and participation in projects tender and selection process; and participation in identification of the state's infrastructure needs both with percentage contribution of 31.30%.

For the financial administrators, that is economists and accountants, the tables reveal the mean contribution of 55.48% for preparation of budget estimate which received the highest ranking. Next to this is technical and cost evaluation of infrastructure projects with percentage contribution of 53.17%; drafting of projects contractual agreement/documentation with contribution of 52.35%; and projects tender and selection process with contribution of 51.19%. Low contributions were obtained in preparation of the macro-economic framework for development of infrastructure sector with contribution of 43.52%; finalization of the draft budget for overall infrastructure sector with 44.25%; and developing of budget strategy to accommodate the infrastructure needs with contribution of 45.10%.

Aggregating the contribution of the construction professionals, the tables reveal contributions that are above average (50.00%) in processes involving projects execution, monitoring and evaluation process (59.06%); projects cost monitoring and control activities (55.58%); preparation of budget estimate for infrastructure projects (55.06%); projects tender and selection process (53.50%); pre-budgetary technical and cost evaluation of infrastructure projects (51.58%); and review and approval of budget estimate for infrastructure projects (51.58%). Their contribution ranges between 15.18% - 49.57% which are below average (50.00%) for other processes which are substantially pre-budgetary process.

These results show that architects are only involved in the budgeting process involving projects execution, monitoring and evaluation process; projects cost monitoring and control activities; project tender and selection process; drafting of projects contractual agreement/documentation; and preparation of budget estimate for infrastructure projects and finalization of the draft budget for overall infrastructure sector. The results also indicate that architects are not adequately involved in identifying the state's infrastructure needs; developing of budget strategy to accommodate the infrastructure needs and preparation of a budget circular which gives guidelines for the preparation of infrastructure sector budgets. From the broad classification of budgeting process for infrastructure development into pre-budgetary and post-budgetary exercise, these results show that architects are only adequately involved in the post-budgetary aspect of the budgeting process for infrastructure development.

Moreover, the results show that the contribution of the town planners has not been adequately incorporated into the budgeting process for infrastructural development. Progressive contributions, however, exist in preparation of budget estimate for infrastructure projects and identification of the state's infrastructure needs. The contribution of the town planners in the post-budgetary processes on average is also revealed to be grossly low.

The results show a relatively low contribution of quantity surveyors in activities involving pre-budgetary processes, that is, preparation of budget circular which gives guidelines for the preparation of infrastructure sector budget; identification of the state's infrastructure needs; finalization of the draft budget for overall infrastructure sector and preparation of the macro-economic framework for infrastructure development. The results, however, show an improved contribution in activities involving post-budgetary process compared to that obtainable in the contribution of town planners. The results also show that, similarly to those obtainable for quantity surveyors, engineers are only adequately involved in projects execution, monitoring and evaluation process; projects cost monitoring and control activities; preparation of budget estimate for infrastructure projects; and participation in review and approval of budget estimate for infrastructure projects; and participation in projects tender and selection process which also represent the post-budgetary activities of the budgeting process.

The involvement of the engineers is, however, below average in budgeting process involving developing of budget strategy to accommodate the infrastructure needs; preparation of implementation plan of budget for infrastructure sector; and preparation of a budget circular which gives guidelines for the preparation of infrastructure sector budgets; and finalization of the draft budget for overall infrastructure sector and preparation of the macroeconomic framework for development of infrastructure sector.

Furthermore, the results show that, similarly to other construction professionals, builders are adequately involved in the post-budgetary process, that is, drafting of project contractual agreement/documentation; tender and selection process; projects execution, monitoring and evaluation process; projects cost monitoring and control activities and project auditing activity. The results also indicate that builders contribute more on the average in the pre-budgetary process in comparison with the contributions obtainable for other construction professionals.

The results show that estate surveyors are involved majorly in activities involving technical and cost evaluation of infrastructure projects; developing of budget strategy to accommodate the infrastructure needs; participation in review and approval of budget estimate for infrastructure projects and preparation of implementation plan of budget for infrastructure sector; preparation of a budget circular which gives guidelines for the preparation of infrastructure sector budgets; and preparation of budget estimate for infrastructure projects. However, unlike those obtainable for architects, quantity surveyors, engineers, the results show that estate surveyors are more adequately involved in the pre-budgetary activities.

While the results show an average contribution of the financial administrators, that is accountants/economists, in both the pre-budgetary and post-budgetary processes, the contribution in processes involving preparation of the macroeconomic framework for development of infrastructure sector and finalization of the draft budget for overall infrastructure were lower than expected. The implication of this may include lack of connectivity between budget size and the infrastructure project and this could negatively affect the implementation of the projects. This study indicated that budgeting process for infrastructure development in Nigeria is more of politics driving than development consideration as asserted by [20].

Comparative evaluation of the contributions of the construction professionals, that is, architects, town planners, quantity surveyors, builders, engineers, estate surveyors in the budgeting process shows the least contribution of town planners and the highest contribution of builders among the professionals. The results also indicate low level contribution of the construction professionals in the activities involving pre-budgetary processes, especially, preparation of the macroeconomic framework for development of infrastructure sector, which received higher rankings for the financial administrators. The results, however, indicate a better involvement of the construction professionals in activities involving post-budgetary activities, that is, drafting of projects contractual agreement/documentation; projects tender and selection process; projects execution, monitoring and evaluation process; and projects cost monitoring and control activities.

While the overall results indicate low level contribution of construction professional in the activities involving pre-budgetary processes, their contribution was revealed averagely adequate in post-budgetary processes. The fact that the contribution of construction professionals in the activities involving pre-budgetary processes were ranked lower indicate that vital professional inputs of these professionals are not adequately incorporated in preparation of the macro-economic framework for development of infrastructure sector; identification the state's infrastructural need; technical and cost evaluation of infrastructure projects; preparation of budget estimate for infrastructure projects; and preparation of implementation plan of budget for infrastructure sector. This may pose negative implication on success of implementation of projects.

The findings established the assertion by [10] and [11] that construction professionals opinion are not adequately incorporated in budgeting process for infrastructure development which suggest the budgeting process is dominated by political executive opinion by which this poses as a significant factor affecting the implementation of infrastructure projects in Nigeria. The result also revealed a low level contribution of the construction professionals in preparation of macro-economic framework for the infrastructure sector. This suggests that this process is either not incorporated in the infrastructure development process in the state or the process is dominated by the political executive input, or better corroborate the reports by the [21] on infrastructure development in Nigeria which identified the absence of macro-economic framework as significant problem of budgeting process and implementation of infrastructure projects in Nigeria.

5.0   IDEA OF 'INFRASTRUCTURE WORKSHOP'

In an attempt to enhance the participation of construction professionals in budgeting for infrastructure development in the state, the idea of 'infrastructure workshop' was suggested in this study.   This was subjected to empirical test of acceptance. The idea is conceived to mean workshop comprising representation of construction professionals, financial administrators and political executives specifically organised in the last quarter of the year to brainstorm on issues relating to infrastructure budget of next fiscal year. These issues among others would include identification of the state infrastructure demand/need, technical evaluation of the projects, cost assessment of the projects to enhance connectivity between the projects and budget, consideration of the state revenue and allocation intended for infrastructure development projects, identification of projects demanding urgent attention that could be incorporated in the budget, and cost/benefit analysis of the projects among others. The response rate obtained on the acceptance of the idea was presented in table 3 below.

In Table 3, respondents were asked to show the level at which the respondents agree to the ideal of infrastructure workshop in enhancing budgetary allocation for infrastructure development in the State. From the table, 38.89 % indicated strong agreement while 33.33% indicated agreement. Eighteen respondents (25.00%) were neither agreed nor disagreed, while only 2.78% considered the ideal a needless approach. This result shows that the idea is very acceptable to the respondents and they were of the opinion that it could significantly improve budgetary allocation to infrastructural development in the state and hence ensure performing budget.

Table 3: Respondents Acceptance of an Infrastructure Workshop

Level of Agreement	Frequency	Percentage (%)
Strongly Agree	28	38.89
Agree	24	33.33
Neutrality	18	25.00
Disagree	2	2.78
Strongly Disagree	0	0.00
Total	72	100.00

Source: Author's Field Survey (2010)

6.0   CONCLUSION

The study indicated adequate contribution in activities involving pre-budgetary technical and cost evaluation of infrastructure projects as well as review and approval of budget estimate for infrastructure projects. This is very good for infrastructural development as the involvement of construction professionals in these activities could enhance better connectivity between budget and the infrastructural projects. The results, however, only indicated a progressive trend in the activities involving pre-budgetary process. While the contributions of financial administrators were ranked higher in the activities involving pre-budgetary process, this is expected. The fact that the contribution of construction professionals were ranked low in the activities involving pre-budgetary processes indicated that vital professional inputs of these professionals are neglected in preparation of the macro-economic framework for development of infrastructure sector; identification the state's infrastructure needs; technical and cost evaluation of infrastructure projects; preparation of budget estimate for infrastructure projects; and preparation of implementation plan of budget for infrastructure sector. This may pose negative implication on the successful of implementation of projects as evident from cases of abandoned and suspended infrastructural projects budgeted for execution within the period under study.

Moreover, the study shows an average contribution in processes involving preparation of the macroeconomic framework for development of infrastructure sector; and finalization of the draft budget for overall infrastructure development were lower than expected for financial administrators. These levels of contribution could be responsible for poor level implementation of infrastructural projects in the state. This study indicated that budgeting process for infrastructure development in Nigeria is more of politics driving than development consideration. The low professionals' contribution obtainable in activities like preparation of the macro-economic framework for development of infrastructure sector and preparation of implementation plan of budget for infrastructure indicated that the political executive' opinion dominates sensitive aspect of infrastructure budgeting process. The dominance of political executive's opinion in the stages like projects' identification and citation may be less significant to projects implementation. From politics view point, the political executive' prerogative on these stages could probably be influenced by their political manifesto and campaign programme.

In summary, the budgeting process for infrastructure development in Nigeria indicated that majority of projects budgeted for execution lack serious technical evaluation and cost assessment as a result of inadequate professional involvement. This could be adduced to be a significant problem of implementation of public financed infrastructure projects in Nigeria. It is thus imperative that a rethinking in budgeting for infrastructure should be considered. It is suggested that the dominance of political influence should be restricted to project identification and citation while more technical stages of the budgeting process should be left to related professionals to make their input. The study recommends the use of infrastructure workshop in addressing budgeting and developmental issues relating to infrastructure in the state. Further studies are also suggested on other factors affecting the implementation of public financed infrastructure in a developing economy.

REFERENCES

[1] Adewoye, O.O., (2007), "Human Capacity Building in Engineering Infrastructure", A paper Presented at the International Conference and Annual General Meeting of the Nigeria Society of Engineer, International Conference Center, Abuja, 4th Dec., 2007.

[2] Opawole, A. Jagboro, G.O. and Babatunde, S.O. (2011a), "Factors Influencing the Implementation of Public Infrastructure in Nigeria", Global Journal of Researches in Engineering, Vol. 11(6), pages: 29-38.

[3] Adegoke O. J., Olaleye A. and Araloyin F.M. (2010), "An Examination Of The Need For Public Private Partnership In The Provision Of Urban Infrastructure In Lagos Metropolis", Ife Journal of Environmental Design and Management, Vol. 4 (1), pages: 76-109.

[4] Awodele, O.A, Ogunsemi, D.R. and Adeniyi, O.O. (2012), "An Appraisal of Private Sector Participation in Infrastructure Development in the Nigerian Construction Industry: Lagos State as a Case Study", Journal of the Nigerian Institute of Quantity Surveyors, Vol. 1 (1), pages: 20-32.

[5] Famakin, I.O., Aje, I.O. and Ogunsemi, D.R. (2012), "Assessment of success factors for joint venture construction projects in Nigeria", Journal of Financial Management of Property and Construction, Vol. 17(2), pages:153 – 165.

[6] Babatunde, S.O., Opawole, A., and Ujaddughe, I.C. (2010), "An Appraisal of Project Procurement Methods in the Nigerian Construction Industry", Civil Engineering Dimension, Vol. 12(1), pages: 1-7.

[7] Bureau of Public Procurement (BPP) (2008), "Procurement Procedures Manual for Public Procurement in Nigeria".

[8] Ayodele, E.O., (2008), "The Roles of Quantity Surveyors in the Economy of Nations", The Journal of The Canadian Institute of Quantity Surveyor, Vol.4, pages:14-19

[9] Gray, C.G., and Larson, E.W. (2003), "Project Management- The Managerial Process", McGraw – Hill Irwin Publisher, New York (NY).

[10] Oforeh E. C., (2006), "The Nigerian Institute of Quantity Surveyors As An Agent Of Economic Development", Paper Presented at 22nd Biennial Conference and General meeting of the Nigerian Institute of Quantity Surveyors, 22nd-25th Nov.,2006 at Channel View Hotel/Conference Centre, Calabar, Cross River State, Nigeria.

[11] Mogbo, T.C. (2001), "the Construction Sector and the Economic Growth of Nigeria, 1981-1995", Journal of the Nigerian Institute of Quantity Surveyors, Vol. 35 (2), pages: 8-13.

[12] Olufidipe, O., (2003), "Government Budgeting in Nigeria: Principle, Policies, and Practices", Obafemi Awolowo University Press Ltd., Ile-Ife, Nigeria.

[13] Wahab, K.A. (2006), "Implementation of Public Procurement Reforms in Nigeria", Presentation at Workshop on World Bank Procurement Procedures, Organized by the Nigeria Economic Summit Group at the Golden Gate Restaurant, Ikoyi , Lagos, 28th September.

[14] Ndukwe, E.C.A., (2007), "Challenge for Rapid Development of Telecommunications and Information Technology Infrastructure for Nigeria".

[15] Familoni, K.A. (2006), "The Role of Economic and Social Infrastructure in Economic Development: A global view", Journal of Economic Perspectives, Vol. 6 (4), pages: 11-32.

[16] Aigbokan, B.E., (2004), "Evaluating Investment of Basic Infrastructure in Nigeria", Journal of African Economies", Vol. 16, (Supplement1), pages: 75-126.

[17] Omotor, D.G. (2004), "An Analysis of Federal Government Expenditure in the Education Sector of Nigeria: implication for National Development", Journal of Social Sciences, Vol. 9 (2), pages: 105-110.

[18] Oyegoke, A.S. (2005), "Infrastructure Project Finance and Execution Development Strategies", Journal of The Nigerian Institute of Quantity Surveyors, Vol.52 (3), pages: 11-19

[19] Olayiwola, L.M. and Adeleye, O.A. (2005), "Rural Infrastructural Development in Nigeria: Between 1960 and 1999-Problems and Challenges", Journal of Social Science, Vol. 11 (2), pages: 91-96.

[20] Robert S.W. and Lynch T. D. (2004), "Public Budgeting in America", 5th Edition. Pearson; Upper Saddle River, New Jersey, pages: 37.

[21] United State Agency for International Development (2005), "Nigeria Budget Process Support Project".

[22] Opawole A., Jagboro G.O. and Babatunde S.O. (2011b), "An Evaluation of the Trend of Budgetary Allocations for Infrastructural Development in Osun State, South-Western, Nigeria", In: Laryea, S., Leiringer, R. and Hughes, W. (Eds). Proceeding of West Africa Built Environment Research (WABER) Conference, 19-21 July 2011, Accra, Ghana, pages: 105-117.

16

COMPARATIVE ANALYSIS OF SANDCRETE HOLLOW BLOCKS AND LATERITE INTERLOCKING BLOCKS AS WALLING ELEMENTS

Akeem Ayinde Raheem[1], Ayodeji Kayode Momoh[2], Aliu Adebayo Soyingbe[2]

[1]Department of Civil Engineering, Ladoke Akintola University of Technology, Ogbomoso, Nigeria.
[2]Department of Building, University of Lagos, Lagos, Nigeria

*Corresponding E-mail : raheemayinde@yahoo.com

ABSTRACT

This study considered the production and testing of sandcrete hollow blocks and laterite interlocking blocks with a view to comparing their physical characteristics and production cost. Some units of sandcrete hollow blocks and laterite interlocking blocks were made using machine vibrated sandcrete block mould and hydraulic interlocking block making machine respectively. The blocks were tested to determine their density and compressive strength. The results obtained from the tests were compared with the specifications of Nigerian Building and Road Research Institute (2006), Nigerian Building Code (2006), and Nigerian Industrial Standards (2000). The results indicated that the compressive strength of 225mm and 150mm sandcrete hollow blocks varies from 1.59 N/mm^2 to 4.25 N/mm^2 and 1.48N/mm^2 to 3.35N/mm^2 respectively, as the curing age increases from 7 to 28 days. For laterite interlocking blocks, the strength varies from 1.70N/mm^2 at 7 days to 5.03N/mm^2 at 28 days. All the blocks produced satisfied the minimum requirements in terms of compressive strength, by all available codes. The cost per square metre of 225mm and 150mm sandcrete hollow blocks are ₦2,808:00 and ₦2,340:00 respectively, while that of laterite interlocking blocks is ₦2,121:20.It was concluded that laterite interlocking blocks have better strength and are cheaper than sandcrete hollow blocks.

Keywords: Sandcrete hollow blocks, Laterite interlocking blocks, Density, Compressive strength

1.0 INTRODUCTION

Walling materials constitute an essential element in housing delivery. It is estimated that it covers about 22% of the total cost of a building. The choice of walling material is a function of cost, availability of material, durability, aesthetics and climatic condition. Barry (1996)[1] defines a wall as a continuous, usually vertical structure of brick, stone, concrete, timber or metal, thin in proportion to its length and height, which encloses and protects a building or serves to divide buildings into compartments or rooms.

The word 'sandcrete' has no standard definition; what most people have done was to define it in a way to suit their own purpose. For the purpose of this study, sandcrete block is a walling unit produced from sand, cement and water. It is widely used in Nigeria as a walling unit. The quality of blocks is a function of the method employed in the production and the properties of the constituent materials. Sandcrete blocks are available for the construction of load bearing and non-load bearing structures [2].

Laterite interlocking blocks is one of the products that Nigerian Building and Road Research Institute (NBRRI) introduced into the construction industry due to the fact that laterite is readily available in Nigeria and that it requires a very small quantity of cement. According to (Mahalinga-Iyer and Williams, 1997) [3], laterite is generally found in tropical and sub-tropical countries. Laterite has been found useful as sub-base or base materials in road construction [3-5].

Nowadays, improved technology induced people to use lateritic interlocking blocks as an alternative for sandcrete blocks in building houses because they do not require cement mortar in bonding the blocks during construction thereby further reducing the building cost [6]. The objective of this study is to compare sandcrete hollow blocks and laterite interlocking blocks in terms of their mechanical properties and cost, with a view to ascertaining which is more applicable in building affordable houses.

2.0   LITERATURE REVIEW

Several studies have been carried out on the use of sandcrete hollow blocks and laterite interlocking blocks as walling units [6-11]. Previous study by the Nigerian Building and Road Research Institute (NBRRI) involved the production of laterite bricks which was used for the construction of a bungalow (Madedor, 1992) [7]. From the study, NBRRI proposed the following minimum specification as requirements for laterite bricks: bulk density of 1810 kg/m^3, water absorption of 12.5%, compressive strength of 1.65 N/mm^2 and durability of 6.9% with maximum cement content fixed at 5%.

Raheem (2006) [6], considered an assessment of the quality of sandcrete blocks produced by LAUTECH Block Industry, an arm of the business ventures of Ladoke Akintola University of Technology, Ogbomoso, Nigeria. The results indicated that compressive strength of 450 x 225 x 225mm (9 inches) blocks increased from 0.54 N/mm^2 at age 3 days to 1.68 N/mm^2 at age 28 days, while that of 450 x 225 x 150mm (6 inches) blocks increased from 0.53 N/mm^2 at age 3 days to 1.59 N/mm^2 at age 28 days. Also, about 60% of the compressive strength at 28 days was developed at day 7 for both 9 and 6 inches blocks.

Raheem et al. (2010a) [10], carried out a comparative study of cement and lime stabilized lateritic interlocking blocks. It was concluded that cement stabilized interlocking blocks were more effective structurally and cheaper than those stabilized with lime. In another study, Raheem et al. (2012)[11], examined the production and testing of lateritic interlocking blocks with laterite samples obtained from Aroje (Ogbomoso North L.G), Olomi (Ogbomoso South L.G), Idioro (Surulere L.G) and Tewure (Orire L.G) of Oyo State, Nigeria. It was concluded that only laterite from Olomi and Idioro that met minimum 7 days requirements are suitable for producing interlocking blocks in the area.

This study carries out a comparative analysis of sandcrete hollow blocks and laterite interlocking blocks by evaluating their mechanical properties and production cost.

3.0   METHODOLOGY

3.1   MATERIALS

All materials used for production of sandcrete hollow blocks and laterite interlocking blocks were obtained locally. Sharp sand was used as fine aggregates and it was made free from deleterious substances by washing. Sieve analysis of the sand was done to determine its grading.

The laterite samples used were air – dried for seven days in a cool, dry place. Air drying was necessary to enhance grinding and sieving of the laterite. After drying, grinding was carried out using a punner and hammer to break the lumps present in the soil. Sieving was then done to remove over size materials from the laterite sample using a wire mesh screen with aperture of about 6mm in diameter as recommended by Oshodi (2004) [12]. Fine materials passing through the sieve were collected for use while those retained were poured away. The liquid limit, plastic limit and plasticity index of the laterite sample were determined in accordance with BS 1377 (1990)[13].

Ordinary Portland cement (Dangote Brand) was used as the binder. The water used was that which is drinkable and free from impurities and it was obtained from a tap in the laboratory.

3.2   PRODUCTION OF SANDRETE HOLLOW BLOCKS

The sandcrete hollow blocks were produced using vibrating block moulding machine with double 150mm (6 inches) moulds and single 225mm (9 inches) mould. Cement and sand were measured in ratio 1:9 by volume batching with the aid of head pans. The materials were then thoroughly mixed together manually until a homogeneous mix with uniform colour was obtained. Water was then added in sufficient quantity to ensure workability of the mixture. The water was judged to be sufficient when a quantity of the mixture pressed between the palms caked without bringing out water [6]. The composite mixture was then introduced into the mould in the block moulding machine and the block vibrated for one minute to ensure adequate compaction as practiced by Raheem (2006) [6]. The green block on wooden pallet was removed from the block moulding machine and placed on the ground for curing. Water was sprinkled on the green blocks, at least twice a day for proper curing for twenty eight days.

3.3   PRODUCTION OF LATERITE INTERLOCKING BLOCKS

The interlocking blocks were produced using hydraulic interlocking block making machine with steel mould of size 230 x 230 x 115mm. The materials used for the production of lateritic interlocking blocks were measured by volume batching. For the 5% cement stabilization adopted, ninety five (95) parts of laterite with five (5) parts of cement i.e. ratio 19:1 (laterite : cement) was used. A four litre plastic container was used as the gauge box. The mixing was done on an impermeable surface made free from all harmful materials which could alter the properties of the mix, by sweeping and brushing or scraping. The measured laterite sample was spread using a shovel to a reasonably large surface area. Cement was then spread evenly on the laterite and the composite material thoroughly mixed with the shovel. The dry mixture was spread again to receive water which was added gradually while mixing, until the optimum moisture content of the mixture was attained.   The optimum moisture content of the mixture was determined by progressively wetting the soil and taking handful of the soil, compressing it firmly in the fist, then allowing it to drop on a hard, flat surface from a height of about 1.10m. When the soil breaks into 4 or 5 parts, the water is considered right (National Building Code, 2006) [14].

The interior of the mould were lubricated so as to prevent the laterite interlocking block from sticking to the sides of the mould and also to give the block a smooth surface. The wet mixture was filled into the mould and then compacted with hydraulic press. After removing the blocks from the machine, they were first allowed to air dry under a shade made with polythene sheet for 24 hours. Thereafter, curing was continued by sprinkling water morning and evening and covering the blocks with polythene sheet for one week to prevent rapid drying out of the blocks which could lead to shrinkage cracking. The blocks were later stacked in rows and columns with maximum of five blocks in a column until, they were ready for compressive strength test.

3.4   TESTING OF THE BLOCKS

Compressive strength and density tests were performed on both sandcrete and interlocking blocks. Compressive strength test was carried out to determine the load bearing capacity of the blocks. The blocks that have attained the ripe ages for compressive strength test of 7, 14, 21 and 28 days were taken from the curing or stacking area to the laboratory, two hours before the test was conducted, to normalize the temperature and to make the block relatively dry or free from moisture. The weight of each block was taken before being placed on the compression testing machine in between metal plates. The block was then crushed and the corresponding failure load recorded. The crushing force was divided by the sectional area of the block to give the compressive strength. The strength value was the average of five specimens.

The density of the block was determined by dividing the weight of the block prior to crushing, with the net volume. The density value was also the average of five specimens.

3.5   COSTING

The unit cost of the sandcrete hollow blocks and laterite interlocking blocks were calculated. Also, the cost per square metre for the sandcrete and interlocking blocks were determined as follows.

3.5.1   UNIT COST OF SANDCRETE HOLLOW BLOCKS

Mix Ratio = 1:9 (that is; one headpan of ordinary Portland cement to nine headpans of sharp sand). This translates to one bag of ordinary Portland cement to eighteen headpans of sand, since there are two headpans in a bag of cement.
The cost of one bag of ordinary Portland cement is ₦1800 as at the time of carrying out the research.

18 head pans of sharp sand @ ₦ 150	= ₦ 2,700
1 bag of cement @ ₦ 1800	= ₦ 1,800
Total	= ₦ 4,500

For 225mm block, ₦ 4,500 produced 25 blocks.

Cost of materials for producing 1 unit of 225mm sandcrete block	= ₦180
Assume 10% for labour	= ₦18
Assume also 20% for plant and others	= ₦36

Total cost incurred in producing one unit of 225mm sandcrete hollow block	= ₦ **234.00**

For 150mm block, ₦ 4,500 produced 30 blocks.

Cost of materials for producing 1 unit of 150mm sandcrete block	= ₦150
Assume 10% for labour	= ₦15
Assume also 20% for plant and others	= ₦30

Total cost incurred in producing one unit of 150mm sandcrete hollow block	= ₦ **195.00**

3.5.2   UNIT COST OF LATERITE INTERLOCKING BLOCKS

Mix Ratio = 1:19 (that is, one part of ordinary Portland cement: nineteen parts of laterite)

A 4 litre Plastic container was used as the gauge in measuring the composition of laterite interlocking block. There are four number of plastic container (each 4litre capacity) in one headpan. Since there are two headpans in one bag of cement, this means that eight number of plastic containers (each 4litre capacity) are contained in one bag of cement.

Cost of buying 1cement bag of laterite = ₦450

The cost of one bag of ordinary Portland cement is ₦1800 as at the time of carrying out the research.

Cost of one, 4litre plastic container of Laterite	$= \dfrac{450}{8}$	= ₦ 56.25
Cost of one, 4litre plastic container of Cement	$= \dfrac{1800}{8}$	= ₦ 225.00

19 parts of laterite @ ₦ 56.25/ part	=	₦ 1,068.75
1 part of cement @ ₦ 225/ part	=	₦ 225.00
Polythene sheet for curing	=	₦ 200.00
Cost of materials used	=	₦ 1493.75

₦ 1493.75 produced 40 blocks

Cost of producing one unit of laterite interlocking block	=	₦ 37.34
Assume 10% for labour	=	₦ 3.73
Assume also 20% for plant and others	=	₦ 7.47

Total cost incurred in producing one unit of laterite interlocking block =   **₦ 48.54**

3.5.3 COST PER SQUARE METRE OF SANDCRETE AND INTERLOCKING BLOCKS

Elevation area of sandcrete block = 0.450m x 0.225m   = 0.1013m^2

Number of sandcrete blocks in one square metre   = $\frac{1.00}{0.1013}$ = 9.87

= Approximately 10 Blocks

Cost of 225mm sandcrete hollow blocks per square metre	= 10 x 234	= ₦2,340.00
Assume 10% for cost of mortar for laying the blocks		= ₦234.00
Assume 10% for labour for laying the blocks		= ₦234.00

TOTAL COST PER SQUARE METRE FOR 225MM SANDCRETE BLOCKS = **₦2,808.00**

Cost of 150mm sandcrete hollow blocks per square metre	= 10 x 195	= ₦1,950.00
Assume 10% for cost of mortar for laying the blocks		= ₦195.00
Assume 10% for labour for laying the blocks		= ₦195.00

TOTAL COST PER SQUARE METRE FOR 150MM SANDCRETE BLOCKS = **₦2,340.00**

Elevation area of laterite interlocking block = 0.230m x 0.115m   = 0.02645m^2

Number of interlocking blocks in one square metre   = $\frac{1.00}{0.02645}$ = 37.81

= Approximately 38 Blocks

Cost of Laterite interlocking blocks per square metre	= 38 x 48.54	= ₦1,844.52
No mortar required		
Assume 15% for labour (more blocks involved)		= ₦276.68

TOTAL COST PER SQUARE METRE FOR LATERITE INTERLOCKING BLOCKS = **₦2,121.20**

4.0   RESULTS AND DISCUSSION

4.1   PHYSICAL PROPERTIES OF SAND

The grading curve for the sharp sand used as fine aggregates is shown in Figure 1.

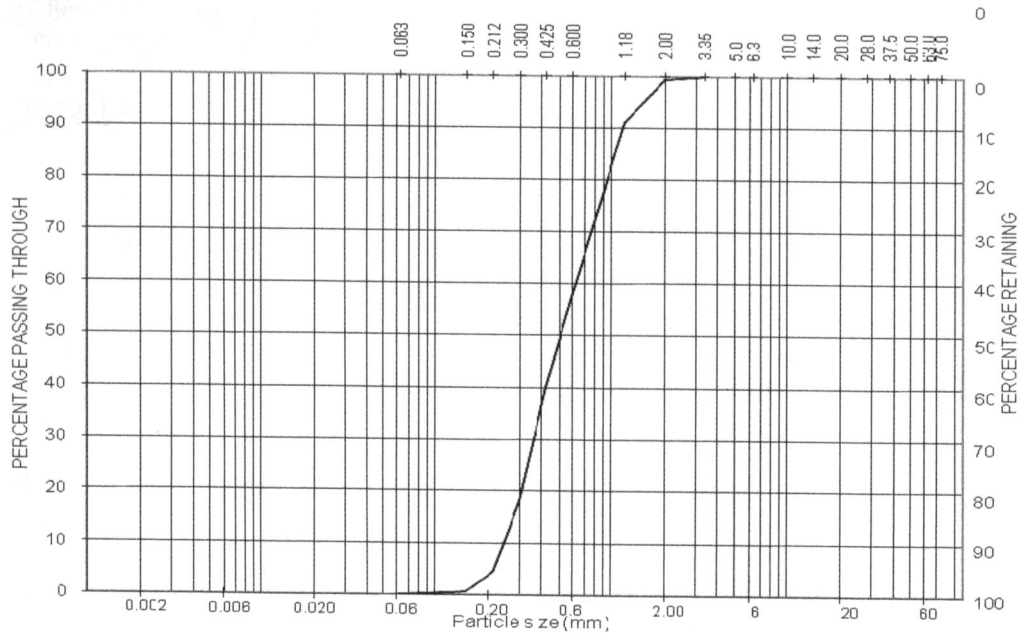

Figure 1: Grading curve for fine aggregates used

It could be observed from the grading curves that the coefficient of uniformity (C_u) and coefficient of curvature (C_c) for the fine aggregates are 2.52 and 0.78 respectively. Thus, the sand can be said to be well graded [15]. The specific gravity of the sand is 2.66.

4.2   PHYSICAL PROPERTIES OF LATERITE

The liquid limit (LL) of the laterite is 45.8% indicating that the soil sample is clayey. The plastic limit (PL) is 17.2% while the plasticity index (PI) is 28.6%. The LL, PL and PI fall within the limits obtained by Raheem et al. (2010b) [16] with value of LL from 42-50%, PL from 10-25%, and PI from 16-39%.

4.3   DENSITY OF BLOCKS

Tables 1 and 2 show the density of 225mm and 150mm sandcrete hollow blocks respectively. It can be observed from Table 1 that the density of 225mm blocks ranges from 2002.21kg/m^3 to 2203.03kg/m^3. These values are slightly higher than those of Raheem (2006) [6] which range from 2073.5 kg/m^3 to 2166.3 kg/m^3. Similar trend was observed for 150mm blocks as indicated in Table 2, with density ranging from 2146.46kg/m^3 to 2209.60kg/m^3 as against 2041.3 kg/m^3 to 2160.9 kg/m^3 in [6]. The higher values of densities recorded in this study may be attributed to the mix ratio used. While the present study uses a mix ratio of 1:9 (cement:sand), Raheem (2006) [6] used a mix ratio of 1:12.

The results of the density of laterite interlocking blocks are presented in Table 3. The density ranges from 6184.21kg/m^3 to 6784.54kg/m^3. It could be observed from the table that the density of laterite interlocking blocks are about three times that of sandcrete hollow blocks. The very high values witnessed are due to the fact that laterite interlocking blocks do not have hollows in them. They are solid blocks; hence, their weights are higher than those of sandcrete hollow blocks.

Table 1: Density of 225mm Sandcrete Hollow Blocks

S/No	Dry Weight (kg)	Dry Density (kg/m^3)	Mean Dry Density (kg/m^3)	Age of Block (day)
1	21.60	2126.25		
2	21.40	2106.56		
3	21.10	2077.03	2086.87	7
4	21.20	2086.87		
5	20.70	2037.65		
1	22.60	2224.68		
2	22.20	2185.31		
3	23.00	2264.06	2203.03	14
4	22.80	2244.37		
5	21.30	2096.71		
1	21.20	2086.87		
2	22.70	2234.53		
3	20.00	1968.75	2002.21	21
4	18.60	1830.93		
5	19.20	1890.00		
1	20.90	2057.34		
2	21.50	2116.40		
3	21.70	2136.09	2102.62	28
4	21.50	2116.40		
5	21.20	2086.87		

Net Volume of block=$1.015875 \times 10^{-2} m^3$

4.4 COMPRESSIVE STRENGTH

Figure 2 shows the results of compressive strength test for sandcrete hollow blocks and laterite interlocking blocks. The result indicated that the compressive strength of 225mm sandcrete hollow blocks varies from 1.59 N/mm^2 at 7 days to 4.25 N/mm^2 at 28 days. For 150mm blocks it varies from 1.48N/mm^2 at 7 days to 3.35N/mm^2 at 28 days. These results are higher than those obtained by Raheem (2006) [6] with values ranging from 1.01 N/mm^2 to 1.68 N/mm^2 and 0.53 N/mm^2 to 1.59 N/mm^2 for 225mm and 150mm sandcrete hollow blocks respectively, during the same period. The higher compressive strength recorded in this study are due to the stronger mix ratio of 1:9 (cement:sand) employed. While only 25 number, 225mm blocks are produced from one bag of cement in the present study, 33 were produced in [6]. Similarly, 30 number, 150mm blocks was produced from a bag of cement in this study as against 42 in [6]. Thus, the reduction in the number of blocks produced per bag of cement resulted in the improved compressive strength. The minimum 28 days compressive strength of 3.40N/mm^2 stipulated by Nigerian Industrial Standard (NIS 87: 2004) [17], for 225mm sandcrete hollow blocks, was satisfied by the blocks produced in this study.

The result of compressive strength test for laterite interlocking blocks as shown in Figure 2 indicated that the compressive strength varies from 1.70N/mm^2 at 7 days to 5.03N/mm^2 at 28 days. The results are higher than those obtained in similar studies by Raheem et al. (2010b) [16] and Raheem et al. (2012) [11], both of which involved manual production of laterite interlocking blocks as against the mechanical means employed in the present study. It can also be observed that at 28 days, the mean compressive strength of laterite interlocking blocks is more than the minimum of 4N/mm^2 stipulated by the Nigerian Building and Road Research Institute (NBRRI) for blocks produced with interlocking block making machine [18]. The Nigerian Building Code (2006) specification of minimum 28 days strength of not less than 2N/mm^2 was also satisfied.

Table 2: Density of 150mm Sandcrete Hollow Blocks

S/No	Dry Weight (kg)	Dry Density (kg/m^3)	Mean Dry Density (kg/m^3)	Age of Block (day)
1	17.80	2247.47		
2	18.30	2310.61		
3	17.50	2209.60	2209.60	7
4	17.00	2146.46		
5	16.90	2133.84		
1	17.70	2234.85		
2	17.30	2184.34		
3	17.60	2222.22	2189.39	14
4	16.80	2121.21		
5	17.30	2184.34		
1	17.10	2159.09		
2	17.10	2159.09		
3	17.00	2146.46	2146.46	21
4	17.20	2171.72		
5	16.60	2095.96		
1	16.70	2108.59		
2	18.10	2285.35		
3	17.40	2196.97	2176.77	28
4	17.60	2222.22		
5	16.40	2070.71		

Net Volume of block=7.92×10^{-3}m^3

Table 3: Density of Laterite Interlocking Blocks

S/No	Dry Weight (kg)	Mean Weight (kg)	Mean Dry Density (kg/m^3)	Age of Block (day)
1	42.75			
2	38.25			
3	43.25	41.25	6784.54	7
4	39.50			
5	42.50			
1	36.50			
2	38.00			
3	37.50	37.60	6184.21	14
4	37.00			
5	39.00			
1	36.50			
2	37.50			
3	36.50	37.90	6233.55	21
4	41.00			
5	38.00			
1	38.25			
2	35.00			
3	37.50	37.65	6192.43	28
4	39.50			
5	38.00			

Volume of interlocking block=6.08 x10^{-3}m^3

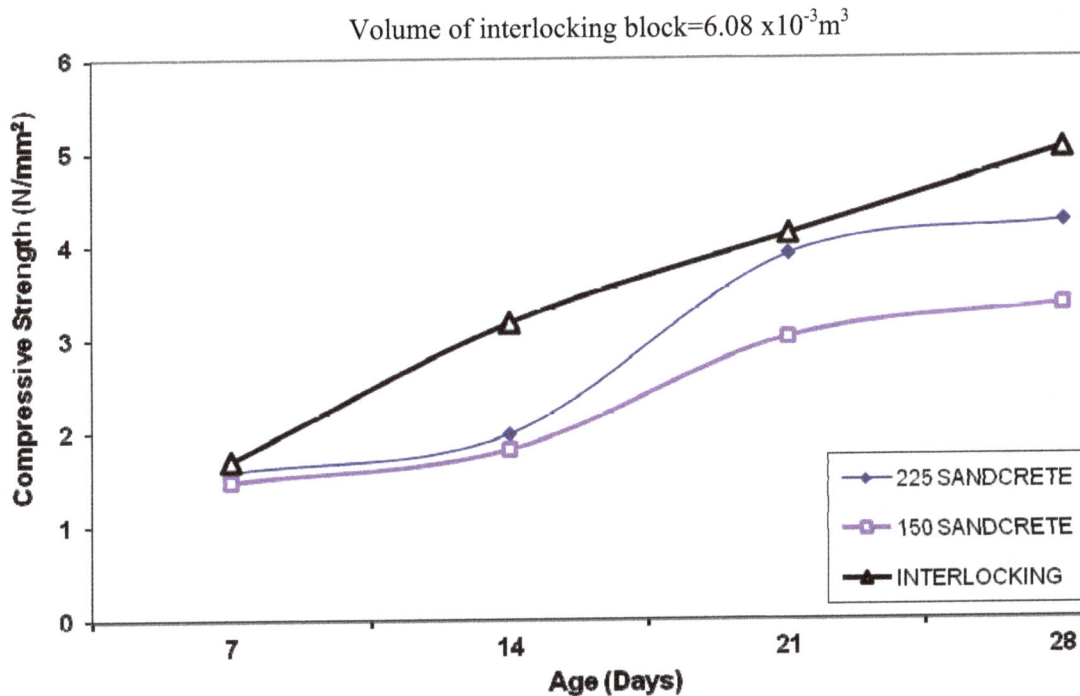

Figure 2: Compressive Strength of Sandcrete and Laterite Interlocking Blocks

4.5   COSTING

The unit cost of 225mm and 150mm sandcrete hollow blocks are ₦234 and ₦195 respectively. This amount is higher than the prevailing price of blocks within the study area which was ₦180 and ₦120 for 225mm and 150mm blocks respectively. The reason for this is that, while only 25 number 225mm sandcrete blocks are produced from one bag of cement in this study, block industries within the study area produced an average of 40 from a bag of cement. Similar trend was observed for 150mm blocks with an average of 50 instead of the 30 blocks produced in this study. The reduction in quantity per bag of cement is responsible for the increase in cost. This is however compensated for by the improvement in quality as witnessed from the higher compressive strength recorded. The unit cost of laterite interlocking blocks is ₦48.54. This very low amount is due to the small quantity of cement (5%) used in producing the blocks since cement is the most costly of all material used. The cost per square metre of 225mm and 150mm sandcrete hollow blocks are ₦2,808:00 and ₦2,340:00 respectively, while that of laterite interlocking blocks is ₦2,121:20. Thus, laterite interlocking block is cheaper both in terms of unit cost and cost per square metre. The interlocking blocks are also aesthetically pleasing and may not require external rendering. It is therefore recommended for building affordable houses.

5.0   CONCLUSION

From the results of the various tests performed and the costing made, the following conclusions can be drawn:

(i)   All the blocks produced satisfied the minimum requirements in terms of compressive strength, by all available codes.

(ii)   Laterite interlocking blocks are denser and stronger than sandcrete hollow blocks.

(iii) Laterite interlocking block is cheaper than sandcrete hollow block, both in terms of unit cost and cost per square metre, hence it is recommended for building affordable houses.

ACKNOWLEDGEMENTS

The authors acknowledge the management and staff of Structures, Concrete and Soil Mechanics laboratories of Civil Engineering Department, University of Lagos, Nigeria; for the opportunity given to perform the various laboratory tests using their facilities.

REFERENCES

[1] Barry, R., (1996),"The Construction of Buildings", 7th edition, Vol 1, England, Blackwell Science.

[2] Hodge, J.C. (1971), "Brick work", 3rd edition, England, Edward Arnold.

[3] Mahalinga-Iyer, U. and Williams, D.J. (1997), "Properties and Performance of Lateritic Soil in Road Pavements", Engineering Geology, Vol 46, pages 71 – 80.

[4] Oladeji O. S. and Raheem A. A. (2002), "Soil Tests for Road Construction", Journal of Science, Engineering and Technology, (JSET), Vol 9 (2), pages 3971-3981.

[5]Raheem, A. A.; Osuolale, O. M.; Onifade, I. and Abubakar, Z. (2010a), "Characterization of Laterite in Ogbomoso, Oyo State, Nigeria", Journal of Engineering Research, Vol 15 (2), pages 73 -79.

[6] Raheem, A. A. (2006), "Comparism of the Quality of Sandcrete Blocks Produced by LAUTECH Block Industry with others within Ogbomoso Township", Science Focus, Vol 11 (1), pages 103-108.

[7] Madedor, A. O. (1992), "The impact of building materials research on low cost housing development in Nigeria", Engineering Focus, Vol 4 (2) pages 37-41.

[8] Agbede, I. O. and Manasseh, J. (2008), "Use of Cement – Sand Admixture in Laterite Brick Production for Low Cost Housing", Leonardo Electronic Journal of Practices and Technologies, Issue 12, pages 163 – 174.

[9] Folagbade, S. O. (1998), "Charateristics of Cement Stabilised Sandy – Lateritic Hollow Blocks", Journal   of Environmental Design and Management, Vol 1 (1&2), pages 67 – 74.

[10] Raheem, A.A.; Bello, O.A. and Makinde, O.A. (2010a). "A Comparative Study of Cement and Lime Stabilized Lateritic Interlocking Blocks". *Pacific Journal of Science and Technology*. 11(2):27-34.

[11] Raheem, A. A.; Falola, O. O. and Adeyeye, K. J. (2012), "Production and Testing of Lateritic Interlocking Blocks", Journal of Construction in Developing Countries (JCDC), Vol.17 (1), In Press.

[12] Oshodi, O. R. (2004), "Techniques of producing and dry stacking interlocking blocks", Paper presented at the Nigerian Building and Road Research Institute (N.B.R.R.I) Workshop on Local Building Materials, Ota, Ogun State, Nigeria.

[13] BS 1377 (1990), "Methods of Test for the Classification of Soil and for the Determination of Basic Physical Properties", London, British Standard Institution.

[14] National Building Code (2006), "Building Regulations", Ohio, LexisNexis Butterworths.

[15] Smith, G. N. and Smith, Ian G. N. (1998), "Elements of Soil Mechanics", 7th edition, London, Blackwell Science.

[16] Raheem, A. A.; Osuolale, O. M.; Onifade, I. and Abubakar, Z. (2010b), "Characterization of Laterite in Ogbomoso, Oyo State, Nigeria", Journal of Engineering Research, Vol 15 (2), pages 73 -79.

[17] NIS 87 (2000), "Nigerian Industrial Standard: Standard for Sandcrete Blocks", Lagos, Nigeria, Standards Organisation of Nigeria.

[18] NBRRI (2006), "NBRRI Interlocking Blockmaking Machine", NBRRI Newsletter, Vol. 1 (1), pages 15 – 17.

17

PROPERTIES OF CONCRETE USING TANJUNG BIN POWER PLANT COAL BOTTOM ASH AND FLY ASH

Abdulhameed Umar Abubakar[1]*, Khairul Salleh Baharudin[2]

[1]Centre for Postgraduate Studies
[2]School of Engineering & Technology Infrastructure
Infrastructure University Kuala Lumpur (formerly KLIUC), Malays

*Corresponding E-mail : abdulhameedabubakar@rocketmail.com

ABSTRACT

Coal combustion by-products (CCPs) have been around since man understood that burning coal generates electricity, and its utilization in concrete production for nearly a century. The concept of sustainable development only reawaken our consciousness to the huge amount of CCPs around us and the need for proper reutilization than the current method of disposal which has severe consequences both to man and the environment. This paper presents the result of utilization of waste from thermal power plants to improve some engineering properties of concrete. Coal bottom ash (CBA) and fly ash were utilized in partial replacement for fine aggregates and cement respectively in the range of 0, 5, 10, 15 & 20% (equal percentages). The results of compressive strength at 7, 28, 56 & 90 days curing are presented because of the pozzolanic reaction; other properties investigated include physical properties, fresh concrete properties and density. The results showed that for a grade 35 concrete with a combination of CBA and fly ash can produce 28 day strength above 30 MPa.

Keywords: *Coal combustion by-products, coal bottom ash, fly ash*

1.0 INTRODUCTION

United Nation Sustainable Development report by [1] defined the concept of sustainable development as "the development that meets the needs of the present without compromising the ability of the future generations to meet their own needs." In engineering practice, sustainability does not stop at the development of new and environment friendly materials for construction purposes, but also the reutilization of materials that were erstwhile considered as waste by-products of industrial processes. According to [2], 'During the last few decades, these "waste" materials have seen a transformation to the status of "by-products" and more recently "products" that are sought for construction and other applications.'

There have been some considerable developments on the research of coal bottom ash in concrete production by [3 – 6] from partial, near total and total replacements of fine aggregates in concrete with some encouraging results. The strength development of coal bottom ash concrete is relatively slow and usually starts beyond 28days [7]; a dramatic increase in performance is noticed beyond this period due to the delay in pozzolanic reaction. The compressive strength, flexural strength and density of bottom ash concrete decreased with an increase in the percentage of bottom ash replacement level [8].

In [9], " when fly ash is used as an admixture in concrete, the early age compressive strength and long term corrosion resisting characteristics of concrete is improved." A recent study by [10] on the development of brick using bottom ash and fly ash showed the 28 days strength of bricks ranged from 4.3-10.96MPa. They concluded that the bricks were comparable to that of normal clay bricks, because it satisfied the minimum strength for Class 1 bricks.

This investigation intends to utilize the special properties of fly ash to improve the properties of coal bottom ash concrete. This will fill in the knowledge gap where currently there is limited or no research done on the combination of the two materials in concrete in Malaysia. Moreover, it has been reported that the properties of CBA varies from one source (power plant) to the other. Partial replacement of CBA and fly ash was done in different percentages, with different mixes using fly ash as replacement for cement and coal bottom ash for fine aggregates. Fresh concrete properties, strength and density were investigated and the result discussed.

2.0   LITERATURE REVIEW

2.1   PHYSICAL PROPERTIES

According to the definition of ACI 116 (as cited in Karim et al., 2011 pp. 4138), fly ash is the finely divided residue resulting from the combustion of ground or powdered coal and which is transported from the firebox through the boiler by the flue gases, known in the UK as pulverized-fuel ash. Sizes may vary from less than $1\mu m$ to more than 80 μm and density of individual particles from less than $1Mg/m^3$ hollow spheres to more than 3 Mg/m^3 (ACI Committee 232). Bottom ash on the other hand has angular particles with a very porous surface texture. It sizes ranges from gravel to fine sand with very low percentages of silt-clay sized particles; the ash is usually a well-graded material, although variation in particle size may be encountered.

The work of [11] showed that Tanjung Bin fly ash exhibit a well graded curve, ranging from fine silt to fine sand sizes, occurring within the range of 0.001mm and 0.6mm; the bottom ash gradation also exhibit a well graded size distribution ranging from fine gravel to fine sand sizes and the majority of the sizes occurred in the range of 0.075mm and 20mm. In [12], working on the same power plant showed that the grain size distribution of fly ash to be well graded from mostly silt to fine sand sizes. Majority of the sizes occurring between 0.001 and 0.06mm, they concluded that Tanjung bin fly ash had more silt particles while bottom ash sizes occurred in a range between 0.03 and 2.00mm.

The specific gravity of fly ash and bottom ash were found to be 2.3 and 1.99 respectively by [11] while [12] arrived at 2.19 and 2.39 respectively for fly ash and bottom ash. They attributed the wide range in specific gravity to two factors (1) Chemical composition with low Iron oxide content resulting in lower specific gravity and vice versa. (2) Presence of hollow fly ash particle or particles of bottom ash with porous or vesicular textures. Fly ash containing a large percentage of hollow particles would have a lower apparent specific gravity than one with mostly solid particles. The apparent specific gravity of bottom ash is also affected by the porosity of its particles. The bottom ash has a higher specific gravity than the fly ash indicating slightly higher Iron oxide content present in the chemical composition of the bottom ash. This can be attributed to the presence of highly porous-popcorn –like bottom ash particles.

2.2   CHEMICAL PROPERTIES

The work of [Muhardi et al., (2010); Fawzan (2010); and Naganathan et al., (2012)] as cited in [13], showed that the major components of the three thermal power plants bottom ash in Malaysia studied were Silica, Alumina & Iron oxide with percentage compositions of 9.78 - 49.4%, 20.75 - 23% & 17 - 37.1% respectively. And that bottom ash used by [Muhardi et al., and Fawzan] is a Class F because the sum of SiO_2 + Al_2O_3 + Fe_2O_3 exceeds 70% and according to ASTM C618 this can be attributed to the use of Bituminous or Anthracite Coal which produce low calcium content. The bottom ash studied by [Naganathan et al.,] is a Class C because the sum is less than 70% but greater than 50%. Class C is generated from the combustion of Lignite or Sub-bituminous coal with a high calcium content. Smaller percentages of potassium, magnesium & sodium are also present in Malaysian power plant bottom ash with traces of barium, manganese

& zinc. BS 3892: Part 1: 1993 specified an SO_3 content of less than 2.5% and a maximum of 5.0% by ASTM C618 and Na_2O alkali of not more than 1.5%.

Table1: Chemical composition of OPC, Bottom ash, Fly ash & requirements.

Chemical contents	Naganathan et al., (2012) Ordinary Portland Cement	Muhardi et al., (2010) Coal bottom ash (%)	Fly ash (%)	Awang et al., (2011) Coal bottom ash (%)	Fly ash (%)	ASTM C618 requirement on the use of fly ash.
SiO_2	21.54	42.7	51.80	46.60	47.10	SiO_2 + Al_2O_3 + $Fe_2O_3 > 70$ Class 'F'
Al_2O_3	5.32	23.0	26.50	26.10	30.00	
Fe_2O_3	3.60	17.0	8.50	12.40	7.34	
CaO	63.60	9.80	4.81	8.31	7.21	>10 Class 'C'
K_2O	-	0.96	3.27	1.34	1.62	-
TiO_2	-	1.64	1.38	1.84	1.83	-
MgO	1.00	1.54	1.10	1.26	1.52	Max 5.0
P_2O_5	-	1.04	0.90	0.62	1.37	-
Na_2O	-	0.29	0.67	0.62	0.72	Max 1.5
SO_3	2.10	1.22	0.60	0.30	0.32	Max 5.0
BaO	-	0.19	0.12	0.13	0.27	-
MnO	-	-	-	-	-	-
ZnO	-	-	-	-	-	-
SrO	-	-	-	0.19		-
CO_2	-	-	-	0.10	0.10	-
Gs	3.15	1.99	2.30	2.39	2.19	-

2.3   MECHANICAL PROPERTIES

Research on the use of coal bottom ash and fly ash in concrete has been carried by partially replacing fly ash with cement and grinding bottom ash particles to smaller sizes as a pozzolana; bottom ash as a partial or total replacement of fine aggregate and as coarse aggregate using large size particles. According to [14], large size (greater than 6mm) bottom ash can be used as coarse aggregate and small size can be used as fine aggregate.

Investigations by [5] showed that it is possible to manufacture lightweight concrete with SSD in the range of 1560 – 1960 kg/m^3 and a 28 day compressive strength in the range of 20 - 40 N/mm^2. The test which was conducted in two series showed that the first series, compressive strength decreased at all ages, but for the second series, the decrease was only observed at 3 day strength. However, there was an increase in strength at 7 & 28days when natural sand and coarse aggregates were replaced

The use of fly ash as an admixture (equal quantity of sand replacement) by [9] to evaluate the compressive strength development and corrosion-resisting characteristics of concrete mixes made with fly ash additions of 0, 20 & 30%, and water – cement ratios of 0.35, 0.40, 0.45 & 0.50 concluded that addition of fly ash as an admixture increases the early age compressive strength and long –term corrosion – resisting characteristics. The work of [15] reported that the strength differential between fly ash concrete specimens and plain concrete specimens became more distinct after 28 days: this is after he replaced fine aggregates with class F fly ash in the range of 0, 10, 20, 30, 40 and 50% at 7, 14, 28, 56, 91 & 365 days. He further stated that compressive strength, splitting tensile strength, flexural strength and modulus of elasticity of fine aggregate (sand) replaced fly ash concrete continued to increase with age for all fly ash percentages.

From the work of [16] "there was a slight decrease in compressive strength with an increase in the bottom ash content" this observation was made when they use class C fly ash and class F bottom ash with the bottom ash as coarse aggregate and class F fly ash with class C." Bakoshi et al., (1998) as cited in Siddique (2003, pp. 540) used bottom ash in amounts of 10 – 40% as replacement for fine aggregate. Test indicate that the compressive strength and tensile strength of bottom ash concrete generally increases with the increase in replacement ratio of fine aggregate and curing age.

3.0   MATERIALS AND METHODS

3.1   MATERIALS

The cement used in this research is Lafarge Phoenix Brand, a brand of Portland Composite cement which satisfied the specification for ordinary Portland cement MS EN 197-1:2007[17]. It is an eco-friendly building material with a minimum of 20% recycled content in its chemical composition and concrete made from this cement releases low levels of hydration heat in the early stages of hydration process.

The CBA and fly ash was obtained from Tanjung Bin power plant in Pontian, Johor. The fly ash was obtained directly from the bottom of the electrostatic precipitator into a sack because of its powdery and dusty nature while the coal bottom ash is transported from the bottom of the boiler to the ash pond as liquid slurry in a 200-250mm diameter pipes. At the lab, it was sprayed in a mixing tray to remove the excess moisture, and then placed in an oven at 105° +/- 5°C. The CBA particles were sieved and the size passing 4.75mm BS Sieve was used in the research. Likewise graded river sand passing the same gradation size was used. Coarse aggregate from crushed stone with a maximum nominal aggregate size of 19mm was used, both the fine and coarse aggregate conform to BS 812: Part 103-1990[18]: Testing Aggregate specification.

3.2   MIX PROPORTION

A control mix containing OPC, natural sand and crushed rock aggregate was designed for a compressive strength of 35MPa at 28 days with a slump range of 25-75mm non-air entrained concrete using ACI Method of mix design. Natural sand was partially replaced with CBA in the range of 5, 10, 15 & 20%. Similar proportion was used for cement replaced with fly ash. The mix proportion is given in table 1 for 9No. 150mm^3 cubes moulds. The water to cementitious ratio (cement + fly ash) was kept constant at 0.48 with many trial mix conducted to ensure that the workability was in the range of the designed slump.

Table 2: Mix design for fly ash, CBA concrete

Quantities	Cement(kg)	Fly ash (kg)	Water (kg)	Fine Agg. (kg)	Bottom ash (kg)	Coarse Agg. (kg)
Per m^3	395.83	-	190.00	617.17	-	1088.00
Control	4.01	-	1.92	8.02	-	12.03
5%FA 5%CBA	3.81	0.20	1.92	7.62	0.40	12.03
10%FA 5%CBA	3.61	0.40	1.92	7.22	0.80	12.03
15%FA 5%CBA	3.41	0.60	1.92	6.81	1.20	12.03
20%FA 5%CBA	3.21	0.80	1.92	6.41	1.60	12.03

3.3   BATCHING AND MIXING

Batching was done by weight using the mix proportion presented in table1. The mixing process was done using mechanical tilting mixer and the procedure was the same as that of the normal weight concrete. Upon emptying the content of the mixer, slump test was conducted in accordance with BS 1881: Part 102: 1983[19] to measure the consistency.

3.4   CASTING AND CURING OF SPECIMEN

Sixty numbers of 150x150x150mm concrete cube samples were casted including the complete batch of 7day samples that were repeated. Each batch mix was made to produce nine cubes to be tested for compressive strength at 7, 28, 56 & 90 days. The curing duration was extended beyond 28 days to study the effect of pozzolanic reaction of CBA which usually manifest after 28days. Density was also determined at the above mentioned curing duration.

The fresh concrete was casted in steel mould in three layers and tampered with a tamping rod, the side of the mould rodded and then compacted on a vibrating table. While on the vibrating table, additional sample was added to fill in the gap created as a result of the vibration. The duration of the vibration usually lasted for 45seconds or when air bubbles appeared on the surface of the concrete, but it should be noted that total absence of entrapped air is not possible. The casted specimen was placed in the laboratory for 24 hrs at 27+/-1°C in accordance to MS 26: Part 1:2009[20] until testing day. Immediately after demoulding, the samples were weighted before immersion in a curing tank.

3.5   DETAILS OF TESTS

Grain size analysis conducted was in accordance to BS 1377-2: 1990[21], more so, an analysis of fine aggregate & bottom ash was conducted with respect to BS 410:2000[22]. Specific gravity was tested using the pycnometer procedure using ASTM D 854-00[23]. The determination of moisture condition of the CBA on oven – dried condition at the time of conducting the experiment was done in accordance to BS EN 1097-6: 2000[24] for a duration of 10, 20 & 30 minutes to ascertain the initial water absorption of the bottom ash. Fresh and hardened concrete properties were determined for the sample prepared; the slump and compacting factor tests were carried out in accordance with BS 1881: Part 102: 1983[19] and BS 1881: Part 103: 1983[25] respectively. The compressive strength test and the density test were carried out in accordance to BS 1881: Part 116: 1983[26] and BS 1881: Part 114: 1983[27] respectively.

4.0   RESULTS AND DISCUSSION

4.1   PHYSICAL PROPERTIES

The sieve analysis of CBA was conducted in accordance to BS 410:2000[22]. The result of the analysis showed that it is distributed from fine gravel to fine sand with a very large percentage of the sand from coarse to medium sand conforming to BS 882:1992[28] requirements.

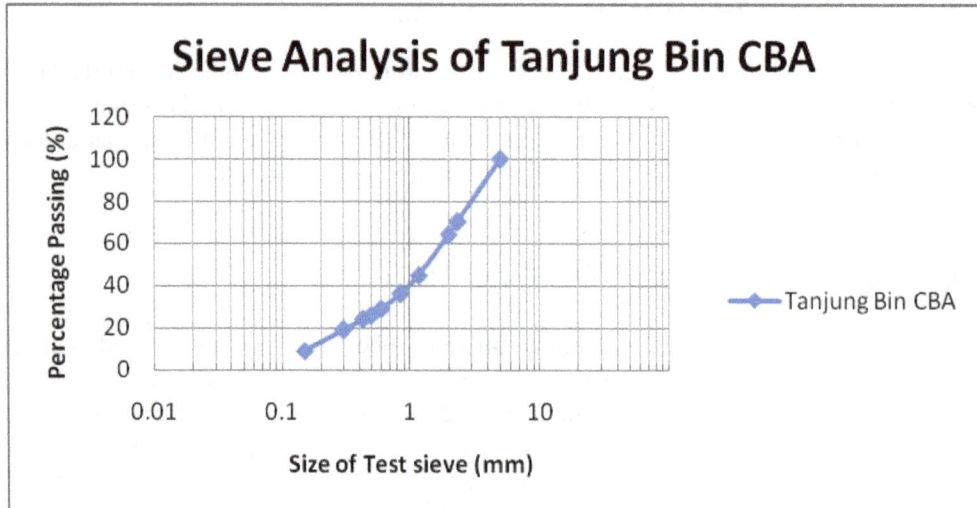

Figure 1: Graphical representation of grain size curve.

The grading analysis of CBA in accordance to BS 410:2000[22] is presented in table 3. The material passing 600µm and 300µm (No. 30 and 50 ASTM) sieve sizes were 29.6 and 18.8% respectively, showing that there were less material in that range, and about 45.2% passing 1.18mm (No. 16) meaning there were more materials larger than 150µm (No. 100) sieve. For the grading limits, Tanjung Bin coal bottom ash satisfies zone 1 and 2 for 600µm, zone 1, 2, 3 and 4 for 300µm (see appendix). The requirements for 1.18mm satisfy for zone 1. BS 882: 1973[29] is based on the percentage passing 600µm.

Table 3: Grain size analysis of CBA from Tanjung Bin

| BS sieve size | Percentage retained (%) | Cumulative percentage retained (%) | Cumulative percentage passing (%) | BS 3797:1990 | | ASTM C 330-89 (%) |
				Grade L1 (%)	Grade L2 (%)	
5.00mm	0.0	0	100	90-100	90-100	85-100
2.36mm	30.0	30.0	70.0	55-100	60-100	-
2.00mm	5.6	35.6	64.4			
1.18mm	19.2	54.8	45.2	35-90	40-80	40-80
850µm	8.8	63.6	36.4			
600µm	6.8	70.4	29.6	20-60	30-60	-
500µm	3.2	73.6	26.4			
425µm	2.0	75.6	24.4			
300µm	5.6	81.2	18.8	10-30	25-40	10-35
150µm	10.0	91.2	8.8	5-19	20-35	5-25
75µm	3.2	94.4	5.6			

Therefore additional limits requirements was satisfied for all the limits of coarse, medium and fines aggregates. Aggregate grading zone 2 and 3 is often described as concreting sand which is derived from BS 882: 1992[28].Tanjung Bin coal bottom ash satisfies the requirement of ACI 213 for fine lightweight aggregate because it has a 100% passing the No.4 sieve size. Grading requirement for lightweight fine aggregate given by BS 3797:1990[30] and ASTM C 330-89 [31] were met but it is always at the lower bound, which means that Tanjung Bin coal bottom ash has a higher percentage of coarse particles.

Figure 2(a) coarse bottom ash sample 5mm and above. **Figure 2(b)** fine bottom ash sample on fine sand passing 5mm sieve size.

The specific gravity of the fly ash and CBA were found to be 2.45 and 1.9 respectively comparatively lower than the specific gravity of cement 3.15 and natural sand 2.6. According to [11], the specific gravity of fly ash has more range with the variety values of and average lower than natural soils due to different chemical content and particle structure; compared to natural sand that is of the same composition. Also, a porous or hollow bottom ash may present a specific gravity as low as 1.6, the porous nature of the bottom ash results in excessive water intake. In [32], it was reported that original bottom ash has a specific gravity of 2.13; ground bottom ash 2.70 and Portland cement 3.14. They attributed the variation in specific gravity between original and ground bottom ash to lower porosity as a result of grinding the ground bottom ash. This result of Tanjung Bin coal ash indicates a low iron oxide and lime content which conforms to the result of [33]. Another factor that might be responsible for the lower specific gravity is the coarse texture of the sand, the result of the sieve analysis indicated that it has a high percentage of particles from 5mm to 1.18mm. The work of [35] showed that bottom ash with low Gs posses a porous and vesicular texture; also [36] reported that porous bottom ash may present low Gs, sometimes as low as 1.6. The state of the material at the time it was utilized also affect the specific gravity, the bottom ash was oven – dried at the time it was tested for specific gravity. Studies have shown that dry bottom ash has a lower specific gravity than saturated bottom ash

The result of the 24hr water absorption of CBA showed that the absorption was 19% which is in line with the specification of ACI 213R[37], that lightweight concrete aggregate generally absorb from 5-20% by weight of dry aggregate depending on the pore structure of the aggregate. This is in contrast to normal weight concrete aggregate which absorb less than 2% moisture.

The CBA aggregate during testing for water absorption rate, was utilized in oven – dried condition and soaked in distilled water for 10, 20 & 30 minutes. It was then surfaced dried before placed in an oven for the equal amount of time it spends in the water. Upon, removal from the oven, it was spread to cool to avoid taking reading with temperature difference. The percentage water absorption rate was 8, 16 & 17 for 10, 20 & 30 minutes respectively. It was observed that the absorption for 10 – 30 minutes was relatively high compared to that of 24 hours water absorption, this is due the fact that former were oven – dried for the equal amount of time they spend in water before it was tested while the latter was dried for 24 hours before it was tested. Some moisture might still be present in the CBA at the time of testing for the 10 – 30 minutes soaking, because the aim is to establish the rate of water absorption over a specified period of time. Another factor that might have warrant the high initial absorption has to do with the particle size of the CBA, the sieve analysis result indicated that it is distributed from fine gravel to fine sand with a large percentage of the sand in the coarse to medium sand gradation. Therefore, the tendency of the particles to have a large porous surface area is very high.

Generally, lightweight aggregates have a high initial absorption of moisture, then the process slow down at a later stage when the inner pores are saturated. According to [38], " for many purposes, the early absorption is the important one and this range from about 5 – 15% of the dry weight after 24 hrs, perhaps 3% to 12% after 30 minutes. The typical data normal aggregates are 0.5% to 2% for 24hrs absorption." Ultimately, the rate of absorption of lightweight aggregate depends on the aggregate type, particle size of the aggregate, the initial condition of the aggregate (whether oven-dried or pre-wetted intentionally or otherwise) as we have seen in the case of coal bottom ash.

4.2   MECHANICAL PROPERTIES

4.2.1   FRESH CONCRETE PROPERTIES

The result of the workability of the fresh concrete was correlated between slump and compacting factor. The results from table 3 indicate that 5% to 15% replacement of CBA/ fly ash exhibit workability within the desired range of 25-75mm slump except for the 20% replacement. BS 1881: Part 102[19] 1983 stipulates that the test method appropriate for medium workability is compacting factor and slump, even though slump is attacked as useless and poor indicator of concrete strength. Its main purpose is to determine variations in uniformity that might occur in a given mix.

Table 4: Workability of the fresh concrete

Sample	Slump measured (mm)	Compacting factor	Workability
Control	55	0.903	Medium
5%	25	0.952	High
10%	40	0.934	Medium
15%	45	0.930	Medium
20%	15	0.882	Low

The increase in the slump value that was observed from 25mm at 5% to between 40 – 45mm at 10 – 15% might be an indication that the water content might have increased, but in this situation, the water - cement ratio was constant at 0.48. According to [39], for a constant workability, the reduction in the water demand of concrete due to fly ash is usually between 5 and 15% by comparism with a Portland cement only mix having the same cementitious material content; the reduction is larger at higher water – cement ratios. Another reason responsible for the increased workability is the "ball-bearing effect" of the fly ash; this is as a result of the spherical shape of the fly ash, the finer particles becomes adsorbed on the surface of cement particles enough to cover the surface of the cement particles; the water demand for a given workability is thus reduced. The reduction in the water demand becomes larger with an increase in the fly ash content up to about 20%. The drop in workability (high – low) was noticed as a result of the percentage increase in bottom ash quantity in the mix. In conclusion, according to Helmuth (1987) [as cited in Neville, 95 pp. 654], 'the action of fly ash, like that of Superplasticizers, on water demand is through dispersion and adsorption of the fly ash particles of Portland cement.' And the lower specific gravity of fly ash compared to that of cement for the same mass means that the volume of fly ash is about 30 per cent higher than that of cement.

Figure 3 (a) workable CBA/FA mix **Figure 3** (b) CBA/FA mix at high percentage replacement ratio.

Bleeding which is a phenomenon whereby water rises to the surface of freshly mixed concrete due to its lower specific gravity among the constituents of the mix. In this research, bleeding was not physically measured, but it was observed that the process was akin to that of conventional concrete, and this could be attributed to the lower percentage of bottom ash-fly ash replacement to have any significant effect. The rise of water to the top of the mould, and subsequent drying did not last more than an hour even at 20% replacement of both bottom ash & fly ash. There are a number of remedies that have been suggested by researchers on how to overcome excessive bleed, one of which is the application of ultra-fine materials and the use of rounded sand with an excess of 15% passing 150µm (Neville, 95 pp.207)[31].

4.2.2 FIGURES AND TABLES

Compressive strength is a very important parameter in accessing the durability of concrete, and one of the methods of measuring strength is the 28 day cube test. In this research, conventional curing in water method was used and the curing duration extended beyond 28 days to study the effect of pozzolanic reaction; fly ash was used to partially replace cement with no other additional admixture was used. At 28 days, the strength of 15% & 20% replacement were 34MPa and 33.6MPa respectively, which is very close to the targeted strength of 35MPa. Figure 4 shows the result of compressive strength at constant water-cement ratio; overall, all the replacement samples were in the range of above 30MPa which is a good sign. Also, at 7 days, the strength of 5, 10, 15 and 20% replacement samples were within 67% to 79% of the 28 days strength.

In general, concrete with proportion of CBA produce lower strength at the early ages. However, the inclusion of fly ash of equal percentage to CBA to replace cement was responsible for the early age strength increment. It was reported by [40] that most of the furnace bottom ash concrete was lower in compressive strength than the control that was manufactured with natural sand up to an age of 28 days for water – cement ratios of 0.45 and 0.55, but most of the FBA concrete was comparable with that of the control concrete at 365 days.

This can be attributed to three factors which are the hydration of the cement and water at the inception. Nucleation effect occurs as a result of smaller particles of fly ash blending with cement paste to accelerate the reaction and form smaller cementing paste. Lastly, packing effect resulted in denser packing of the material in which fine particles that were not reacted fill in the voids spaces present.

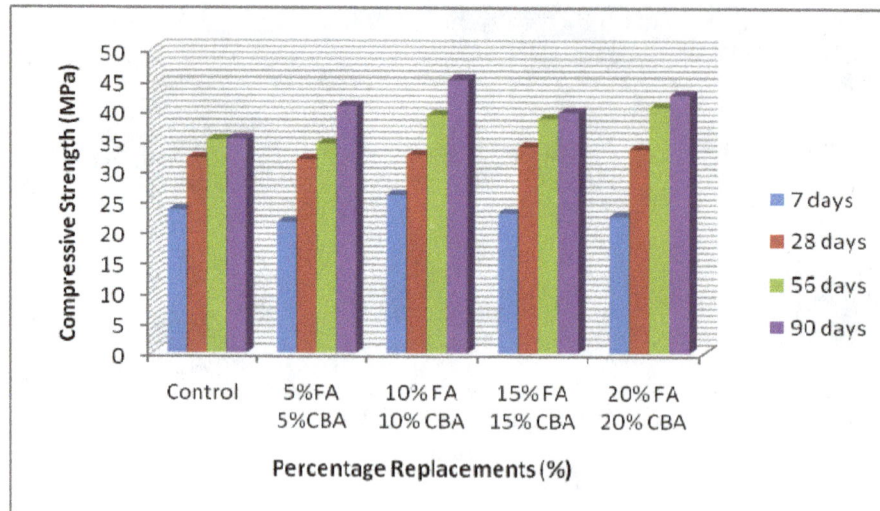

Figure 4: Result of compressive strength for CBA/FA concrete.

In order to have a good, durable concrete, it is good to have gradation of aggregates from large to small, and enough cementing material so as to bind properly. Mechanically, tiny fly ash particles fill voids in concrete due to its hard and round nature, and it also produces a "ball bearing effect" that allows concrete to flow easily into voids.

When the curing age was extended beyond 28 days, all the samples achieved strength above the targeted strength. This can be attributed to the Pozzolanic reaction that normally manifest at a later ages in which $Ca(OH)_2$ reacting with SiO_2 and Al_2O_3 to form C-S-H gel; two materials with pozzolanic properties are present, bottom ash which has been reported to have pozzolanic properties [32] and fly ash, the combination of whom in the presence of lime form cementitious compounds as shown below:

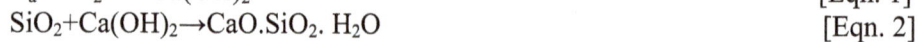

$$C_aO + H_2O \rightarrow Ca(OH)_2 \qquad \qquad \text{[Eqn. 1]}$$
$$SiO_2 + Ca(OH)_2 \rightarrow CaO.SiO_2. H_2O \qquad \qquad \text{[Eqn. 2]}$$

The result of the compressive strength with respect to the percentage increase in strength of the control concrete is presented in table 4. It can be seen that at the age of 90 days curing period, all the replacement samples had strength greater than that of the control sample. This can be attributed to the Pozzolanic reactions that normally manifest at later ages to form C-S-H gel especially in bottom ash; this is a result of larger particles of bottom ash reacting with calcium hydroxide unlike in fly ash that the particle sizes are smaller and reaction is almost immediate.

Table 5: Percentage increase in compressive strength of Concrete samples

Compressive strength (MPa) – Percentage Compressive strength (%)								
Sample marking	7 days		28 days		56 days		90 days	
	MPa	%	MPa	%	MPa	%	MPa	%
Control	23.5	100	32.1	100	35.1	100	35.3	100
5%FA 5%CBA	21.5	91	31.9	99	34.5	98	40.7	115
10%FA 10% CBA	26.0	111	32.6	102	39.3	112	45.1	128
15%FA 15%CBA	23.0	98	34.0	106	38.7	110	39.7	112
20%FA 20%CBA	22.4	95	33.6	105	40.6	116	42.6	121

It should also be noted that keeping the water to cementitious ratio constant at 0.48 was responsible for the strength gain of the concrete, because at higher water-cement ratio, the strength of concrete reduces drastically and vice versa.

The result of the compressive strength was in contrast to what has been reported by many researchers in the literature, it is a widely held believe that the strength development of bottom ash concrete is initially slower at the beginning [3]. However, the strength gain follows the pattern of the control concrete, and this can be attributed to the addition of the fly ash in equal percentage of the bottom ash and also keeping the water - cementitious ratio low because at higher w/c ratio, the strength of concrete decreases drastically. This gives hope that with appropriate replacement ratio, bottom & fly ashes can be utilized in concrete without short and long term durability effects.

The density of the resulting concrete samples with replacement of equal percentage of bottom ash fluctuates at smaller percentages in all curing ages. As the percentage replacements began to increase from 15% to 20%, a decline of density was observed due to lower specific gravity of both fly ash and CBA which conform to the result of [5].

Table 6: Air-dried Density of CBA/FA concrete at 28 days

Sample	Control	5%	10%	15%	20%
(kg/m³)	2326	2313	2323	2331	2295

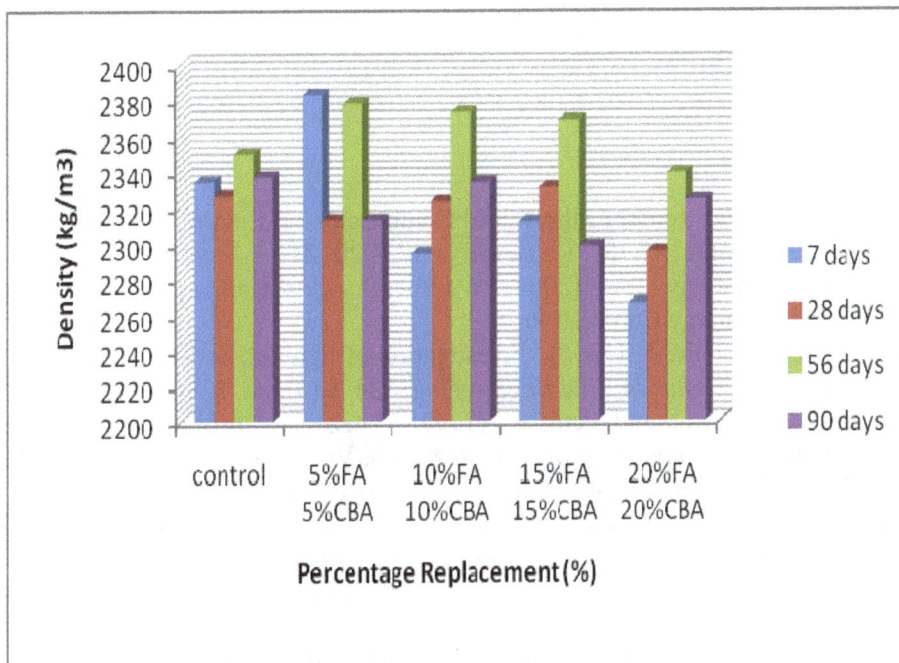

Figure 5: Density of CBA/fly ash concrete at constant w/c ratio.

The density was higher at lower water/cement ratio, because of the cohesiveness of the mix and the bond strength between the cement and aggregate unlike at higher w/c ratio where a considerable amount of water was required to attain the desired workability. Another important factor that might have caused the lower density was the low specific gravity of both bottom ash and fly ash with respect to that of sand and cement, as well as the porosity. The work of [5] reported that the use of lytag, PFA and bottom ash produce a low density at 28 days SSD, and a further decline of density was noticed when they replaced 30% of OPC with fly ash. Therefore, the density of bottom ash – fly ash concrete decreased with an increase of replacement ratio, slightly at the beginning but at higher replacement, drastic decreased is noticed. This is in conformity with Lydon and Balendran (1986) that "the density of concrete increases with the increase in the density of aggregate" (as cited in Neville, 1995, pp 419)[39].

5.0   CONCLUSIONS

The result of the grain size analysis indicated that Tanjung Bin coal bottom ash is distributed from fine gravel to fine sand with a higher percentage of coarse sand particles. The lower specific gravity of fly ash & bottom ash is as a result of the chemical composition which is lower in lime & alumina content (2.45 & 1.9). Tanjung Bin bottom ash aggregate absorbed 19% by weight of dry aggregate in contrast to just 2% of normal weight aggregate and it also indicates that there is a direct relationship between the rate of absorption & time. The workability of the fresh concrete measured in terms of slump and compacting factor decreased as the percentage replacement increases. An increase in curing duration is required to attain maximum compressive strength; the targeted strength was achieved at 56 days curing period for all the percentage replacements. The air-dried density of the concrete showed a marked decline due to low specific gravity of both fly ash and CBA. In conclusion, the use of coal bottom ash and fly ash in concrete has the potential to produce long term durable and good strength concrete. Nevertheless, further research is required on other properties of the CBA/ fly ash concrete.

ACKNOWLEDGEMENT

The authors will like to give gratitude to almighty Allah for giving them the gift of life, and the management of Malakoff Corporation the operators of Tanjung Bin power plant for facilitating the use of their plant waste. The effort of the technical staff of Ikram Material Testing Laboratory cannot be over emphasized for their valuable assistance.

REFERENCES

[1] United Nations (1987) "Report of the World Commission on Environment and Development" General Assembly Resolution 42/187. Retrieved on 6/24/2012

[2] Ramme, B. W. & Tharaniyil, M.P (2004) "We Energies Coal Combustion Products Utilization Handbook" Wisconsin Energy Corporation.

[3] Ghafoori, N. and Bucholc, J., (1996) "Investigation of lignite-based bottom ash for structural Concrete". *Journal of Materials in Civil Engineering*, Vol. **8** (3), pp. 128–137.

[4] Cheriaf, M., Rocha, J.C. and Pera, J., (1999) "Pozzolanic properties of pulverized coal Combustion bottom ash. *Cement and Concrete Research*, Vol. 29 Issue 9 pp.1387-1391

[5] Bai, Y., Ibrahim, R. and Basheer, P.A.M., (2010) "Properties of lightweight concrete Manufactured with fly ash, furnace bottom ash and Lytag". International Workshop on Sustainable Development and Concrete Technology, Pp. 77-88.

[6] Fawzan, A.A., (2010) "Bottom Ash as a Sand replacement in concrete mix". Unpublished B.Eng. Thesis KLIUC, Malaysia.

[7] Kula, I.et al., (2001) "Effects of colemanite waste, coal bottom ash and fly ash on the properties of cement" *Cement and Concrete Research*. Vol. 31 pp. 491-494.

[8] Arenas, C.G., Marrero, M., Leiva, C., Solis-Guzman, J., Arenas, L.F.V. (2011) "High fire resistance in blocks containing coal combustion fly ashes and bottom Ash. *Waste Management*, Vol. 31, Issue 2 pp. 246-252.

[9] Maslehuddin, M., (1989) "Effect of sand replacement on the early – age strength gain and long – term corrosion – resisting characteristics of fly ash concrete" *ACI mater. J.* 86(1) pp. 58 – 62.

[10] Naganathan, S., Subramaniam, N. & Mustapha, K.N. (2012) "Development of Brick using Thermal Power Plant Bottom Ash and Fly ash". *Asian Journal of Civil Engineering (Building and Housing)*, Vol. 13 No. 1 pp. 275-287.

[11] Muhardi et al., (2010) "Engineering characteristics of Tanjung Bin coal ash" EJGE Vol. 15, Bund. K. pp. 1117 – 1129

[12]Awang, A., Marto, A. and Makhtar, A.M. (2011) "Geotechnical properties of Tanjung Bin coal ash mixtures for backfill materials in embankment construction." EJGE Vol. 16 pp. 1515 – 1531.

[13] Abubakar, A.U. and Baharudin, K.S. (2012) "Potential use of Malaysian thermal power plants coal bottom ash in construction" *International Journal of Sustainable Construction Engineering & Technology* Vol. 3 Issue 2 pp. 25 - 37

[14]Siddique, R. (2010)"Utilization of coal combustion by-products in sustainable construction materials. *Resources, conservation and recycling*. Vol. 54. Issue 12. 1060-1066.

[15] Siddique, R. (2003) "Effect of fine aggregate replacement with Class F fly ash on the mechanical properties of concrete." *Cement and Concrete Research* 33 pp. 539 – 547.

[16] Wei, L., Naik, T. R. and Golden, D. M., (1994) "Construction Materials Made with Coal Combustion By-Products" *Cement, Concrete & Aggregate.* American Society For Testing and Materials.

[17] MS EN 197-1: 2007 "Cement-Part 1: Composition, Specifications and Conformity Criteria for common Cements." Department of Standards Malaysia.

[18] BS 812: Part 103- 1990 "Methods for determination of particle size distribution" British Standard Institution.

[19] BS 1881: Part 102: 1983 "Method for determination of slump" British Standard Institution.

[20] MS 26: Part 1: 2009 "Testing of Concrete: Fresh Concrete" Department of Standards Malaysia.

[21] BS 1377-2:1990 "Method of testing for soils for civil engineering purposes." Classification Tests. British Standard Institution.

[22] BS 410:2000 "Test Sieves: Technical Requirements and Testing." British Standard Institution.

[23] ASTM D 854-00 "Standard Test Method for Specific Gravity of Soil Solids by Water Pycnometer." American Society for Testing and Materials. Annual book of ASTM Standards. V. 04. 02 Constructions. Philadelphia, USA.

[24] BS EN 1097-6:2000 "Test for Mechanical Properties of Aggregates. Determination of Particle Density and Water Absorption. British Standard Institution.

[25] BS 1881: Part 103: 1983 "Method for determination of compacting factor" British Standard Institution.

[26] BS 1881: Part 116:1983 "Method for determination of compressive strength of concrete cubes" British Standard Institution.

[27] BS 1881: Part 114:1983 "Methods for determination of density of hardened concrete" British Standard Institution.

[28] BS 882: 1992 "Specifications for aggregate from natural sources for concrete." London. British Standard Institution.

[29] BS 882: 1973 "Specifications for aggregate from natural sources for concrete." London. British Standard Institution.

[30] BS 3797: 1990 "Specification for lightweight aggregate for masonry units and structural concrete." British Standard Institution.

[31] ASTM C330: Standard specification for lightweight aggregate for structural concretes. American Society for Testing and Materials. Annual book of ASTM Standards. V. 04. 02 Construction. Philadelphia, USA.

[32] Jaturapitakkul, C. and Cheerarot, R., (2003) "Development of bottom ash as pozzolanic material". *Journal of Materials in Civil Engineering*, Vol.15 (1), pp. 48-53.

[33] Abdul Talib, N.R., (2010) "Engineering characteristics of Bottom ash from power plants in Malaysia". Unpublished B.Eng. Thesis. UTM.

[34] ASTM C618 (2006) "Standard specification for coal fly ash and raw calcined natural pozzolan for use in Portland cement concrete." American Society for Testing and Materials. Annual book of ASTM Standards. V. 04. 02 Constructions. Philadelphia, USA.

[35] Lovell, C.W., Huang, W.H. and Lovell, J.E., (1991) "Bottom ash as highway material" Presented at the 70th Annual Meeting of the Transportation Research Board, Washington, D.C.

[36] Kim, B.J., Yoon, S.M. and Balunaini, U., (2006) "Determination of ash mixture properties and construction of test embankment-part A." Journal of Transportation Research Program, Final Report, FHWA/IN/JTRP-2006/24. Purdue University, W. Lafayette, Indiana.

[37] ACI Committee Report 213R. Guide for Structural Lightweight Aggregate Concrete. American Concrete Institute, Detroit. USA

[38] Zulkarnain, F., and Ramli, M., (2008) "Durability performance of lightweight aggregate concrete for housing construction" Proceedings from ICBEDC '08: The 2nd International Conference on Built Environment in Developing Countries. Pp. 541-551.

[39] Neville, A.M., (1995) "Properties of Concrete" Pearson education limited.

[40]Bai, Y. and Basheer, P.A.M. (2003) "Influence of furnace bottom ash on properties of concrete" Proceedings of the ICE – Structures and Buildings. Pp. 85 – 92.

[41]Karim, M.R. et al., (2011) "Strength development of mortar and concrete containing fly ash: A review" *Intern. Jour. Of the Phys. Sci.,* Vol. 6 (17), pp. 4137 – 4153.

[42]ACI Committee 116 (1992). Cement and Concrete Terminology. In ACI Manual of Concrete practice, part 1. American Concrete Institute. Pp. 1 – 68.

Appendix: Grading limits of fine aggregate according to BS 882

BS sieve size	Percentage by weight passing BS sieve (%)							
	Overall limit	Additional limits			Zone 1	Zone 2	Zone 3	Zone 4
		C	M	F				
10 mm	100	-	-	-	100	100	100	100
5 mm	89-100	-	-	-	90-100	90 – 100	90-100	95-100
2.36 mm	60-100	60-100	65-100	80-100	60-95	75-100	85-100	95-100
1.18 mm	30-100	30-90	45-100	70-100	30-70	55-90	75-100	90-100
600 µm	15-100	15-54	25-80	55-100	15-34	35-59	60-79	80-100
300 µm	5-70	5-40	5-48	5-70	5-20	8-30	12-40	15-50
150 µm	0	-	-	-	0-10	0-10	0-10	0-15

Permissions

The contributors of this book come from diverse backgrounds, making this book a truly international effort. This book will bring forth new frontiers with its revolutionizing research information and detailed analysis of the nascent developments around the world.

We would like to thank all the contributing authors for lending their expertise to make the book truly unique. They have played a crucial role in the development of this book. Without their invaluable contributions this book wouldn't have been possible. They have made vital efforts to compile up to date information on the varied aspects of this subject to make this book a valuable addition to the collection of many professionals and students.

This book was conceptualized with the vision of imparting up-to-date information and advanced data in this field. To ensure the same, a matchless editorial board was set up. Every individual on the board went through rigorous rounds of assessment to prove their worth. After which they invested a large part of their time researching and compiling the most relevant data for our readers.

The editorial board has been involved in producing this book since its inception. They have spent rigorous hours researching and exploring the diverse topics which have resulted in the successful publishing of this book. They have passed on their knowledge of decades through this book. To expedite this challenging task, the publisher supported the team at every step. A small team of assistant editors was also appointed to further simplify the editing procedure and attain best results for the readers.

Apart from the editorial board, the designing team has also invested a significant amount of their time in understanding the subject and creating the most relevant covers. They scrutinized every image to scout for the most suitable representation of the subject and create an appropriate cover for the book.

The publishing team has been an ardent support to the editorial, designing and production team. Their endless efforts to recruit the best for this project, has resulted in the accomplishment of this book. They are a veteran in the field of academics and their pool of knowledge is as vast as their experience in printing. Their expertise and guidance has proved useful at every step. Their uncompromising quality standards have made this book an exceptional effort. Their encouragement from time to time has been an inspiration for everyone.

The publisher and the editorial board hope that this book will prove to be a valuable piece of knowledge for researchers, students, practitioners and scholars across the globe.

List of Contributors

Faremi Julius Olajide
Department of Building, Faculty of Environmental Sciences, University of Lagos, Akoka, Lagos,Nigeria

Adenuga Olumide Afolarin
Department of Building, Faculty of Environmental Sciences, University of Lagos, Akoka, Lagos,Nigeria

Yen Lei Voo
Dura Technology Sdn. Bhd., Malaysia

Behzad Nematollahi
Civil Engineering Department, Universiti Putra Malaysia (UPM), Serdang, Malaysia

Abu Bakar Bin Mohamed Said
Jabatan Kerja Raya, Kinta Daerah, Perak, Malaysia

Balamurugan A Gopal
Jabatan Kerja Raya, Kinta Daerah, Perak, Malaysia

Tet Shun Yee
TS Yee and Associates, Perak, Malaysia

Hamid Pesaran Behbahani
Faculty of Civil Engineering, Universiti Teknologi Malaysia (UTM), Skudai, Malaysia

Behzad Nematollahi
Civil Engineering Department, Universiti Putra Malaysia (UPM), Serdang, Malaysia

Abdul Rahman Mohd. Sam
Faculty of Civil Engineering, Universiti Teknologi Malaysia (UTM), Skudai, Malaysia

F. C. Lai
Regional Technology Support Centre, Sika Kimia Sdn Bhd, Malaysia

Yen Lei Voo
Dura Technology Sdn Bhd, Malaysia

Patrick C. Augustin
Perunding Faisal, Abraham dan Augustin Sdn Bhd, Malaysia

Thomas A. J. Thamboe
Endeavour Consult Sdn Bhd, Negeri Sembilan, Malaysia

Abdulhameed Umar Abubakar
Research Student, Centre for Postgraduate Studies, Kuala Lumpur Infrastructure University College

Khairul Salleh Baharudin
Assoc. Prof. Dept. of Civil Engineering & Infrastructure Technology, KLIUC, Selangor, Malaysia

Chai Teck Jung
Department of Civil Engineering, Politeknik Kuching Sarawak, Km 22, Jalan Matang, 93050 Kuching, Sarawak

Lee Yee Loon
Department of Structure & Material Engineering, Faculty of Civil and Environmental Engineering, Universiti Tun Hussein Onn Malaysia, 86400 Parit Raja, Batu Pahat,Johor

Tang Hing Kwong
Department of Civil Engineering, Politeknik Kuching Sarawak, Km 22, Jalan Matang, 93050 Kuching, Sarawak

Koh Heng Boon
Department of Structure & Material Engineering, Faculty of Civil and Environmental Engineering, Universiti Tun Hussein Onn Malaysia, 86400 Parit Raja, Batu Pahat,Johor

U. N .Okonkwo
Department of Civil Engineering, Michael Okpara University of Agriculture Umudike, PMB 7267,Umuahia, Abia State, Nigeria

I. C.Odiong
Department of Civil Engineering, Michael Okpara University of Agriculture Umudike, PMB 7267,Umuahia, Abia State, Nigeria

E. E..Akpabio
Department of Civil Engineering, Michael Okpara University of Agriculture Umudike, PMB 7267,Umuahia, Abia State, Nigeria

Rudi Setiadji
Experimental Station of Building Materials, Research Institute for Human Settlements

Andriati Amir Husin
Experimental Station of Building Materials, Research Institute for Human Settlements

Pejman Rezakhani
Department of Architecture and Civil Engineering, Kyungpook National University, Korea

M.R. Alam
Faculty of Engineering and Built Environment, Universiti Kebangsaan Malaysia, Selangor, Malaysia

M.F.M. Zain
Faculty of Engineering and Built Environment, Universiti Kebangsaan Malaysia, Selangor, Malaysia

A.B.M.A. Kaish
Faculty of Engineering and Built Environment, Universiti Kebangsaan Malaysia, Selangor, Malaysia

Kulkarni Swati Ajay
Applied Mechanics Department, SVNIT, Surat, India

R Vesmawala Gaurang
Applied Mechanics Department, SVNIT, Surat, India

AeslinaBinti Abdul Kadir
Faculty of Civil and Environmental Engineering, UniversitiTun Hussein Onn Malaysia, Johor, Malaysia

Norzila Binti Othman
Faculty of Civil and Environmental Engineering, UniversitiTun Hussein Onn Malaysia, Johor, Malaysia

NurulAzimahBinti M.Azmi
Faculty of Civil and Environmental Engineering, UniversitiTun Hussein Onn Malaysia, Johor, Malaysia

Akaninyene A. Umoh
Building Department, University of Uyo,Uyo, Nigeria

K. O. Olusola
Building Department, Obafemi Awolowo University, Ile-Ife, Nigeria

Ranjit K. Nath
Dept. of Chemical & Process Engineering, Universiti Kebangsaan Malaysia(UKM), Bangi, Malaysia

M. F.M .Zain
Dept. of Civil and Structural Engineering, Universiti Kebangsaan Malaysia(UKM), Bangi, Malaysia

Md. Rabiul Alam
Dept. of Civil and Structural Engineering, Universiti Kebangsaan Malaysia(UKM), Bangi, Malaysia

Abdul Amir H. Kadhum
Dept. of Chemical & Process Engineering, Universiti Kebangsaan Malaysia(UKM), Bangi, Malaysia

A.B.M.A. Kaish
Dept. of Civil and Structural Engineering, Universiti Kebangsaan Malaysia(UKM), Bangi, Malaysia

A.Opawole
Department of Quantity Surveying, Obafemi Awolowo University, Ile-Ife, Nigeria

G.O. Jagboro
Department of Quantity Surveying, Obafemi Awolowo University, Ile-Ife, Nigeria

O. Babalola
Department of Quantity Surveying, Obafemi Awolowo University, Ile-Ife, Nigeria

S.O Babatunde
Department of Quantity Surveying, Obafemi Awolowo University, Ile-Ife, Nigeria

Akeem Ayinde Raheem
Department of Civil Engineering, Ladoke Akintola University of Technology, Ogbomoso, Nigeria

Ayodeji Kayode Momoh
Department of Building, University of Lagos, Lagos, Nigeria

Aliu Adebayo Soyingbe
Department of Building, University of Lagos, Lagos, Nigeria

Abdulhameed Umar Abubakar
Centre for Postgraduate Studies

Khairul Salleh Baharudin
School of Engineering & Technology Infrastructure Infrastructure University Kuala Lumpur (formerly KLIUC), Malays